Heat Pump Planning Handbook

The *Heat Pump Planning Handbook* contains practical information and guidance on the design, planning and selection of heat pump systems, allowing engineers, designers, architects and construction specialists to compare a number of different systems and options.

Including detailed descriptions of components and their functions and reflecting the current state of technology, this guide contains sample tasks and solutions, as well as new model calculations and planning evaluations. It also covers economic factors and alternative energy sources, which are essential at a time of rising heat costs.

Topics include:

- ecological and economic aspects;
- introduction to refrigeration;
- water heat pump systems;
- configuration of all necessary components; and
- planning examples (problems and solutions).

Jürgen Bonin is the founder of Umwelt & Technik, Germany.

Heat Pump Planning Handbook

Jürgen Bonin

Translated from the German
by Joanna Scudamore-Trezek

Routledge
Taylor & Francis Group

LONDON AND NEW YORK

Second edition of *Handbuch Wärmepumpen* published in 2012
by Beuth Verlag GmbH

The English translation of the publication "Handbuch Wärmepumpen"
was produced in 2015 with permission of Beuth Verlag GmbH, Berlin.

This edition published 2015
by Routledge
4 Park Square, Milton Park, Abingdon, Oxon OX14 4RN

and by Routledge
605 Third Avenue, New York, NY 10017

Routledge is an imprint of the Taylor & Francis Group, an informa business

© 2015 Published by agreement with Beuth Verlag GmbH, Am DIN-Platz,
Burggrafenstraße 6, 10787 Berlin, Germany

British Library Cataloguing-in-Publication Data
A catalogue record for this book is available from the British Library

Library of Congress Cataloging-in-Publication Data
Bonin, Jürgen.
    Heat pump planning handbook / Jürgen Bonin.
    pages cm
    1. Heat pumps – Installation – Planning – Handbooks, manuals, etc.
    I. Title.
    TH7638.B66 2015
    621.5′63 – dc23
    2014043192

ISBN: 978-1-138-78482-6 (pbk)
ISBN: 978-1-315-70858-4 (ebk)

Typeset in Goudy
by Florence Production Ltd, Stoodleigh, Devon, UK

Contents

Portrait of the author   vii
Foreword   ix

1 Introduction   1

2 Observations on ecology and economics   2

3 What is a heat pump and how does it work?   10

4 Heat pump output   55

5 Planning a heat pump system   78

6 Groundwater protection   158

7 Hydraulics   166

8 Heat pump system planning guidelines   199

9 Different types of heat pump system   202

10 Economic considerations   209

11 Laws and institutions for the protection of people and
    the environment   217

12 Starting up heat pump systems   222

13 Common heat pump errors   226

14 Concluding observations and outlook   231

15 Exercises   233

16 Questions   257

17 Examples of heat pump systems   259

18 Solutions to the exercises   264

19 Answers   318

    Index   324

Portrait of the author

The author was born in 1957 and studied at the University of Wuppertal.

In 1994, he founded Umwelt & Technik, selling water treatment systems and ecological housebuilding equipment. He built his own home in 1998 based entirely on ecological principles, and it is here that he works. It is a half-timbered, low-energy house, which, of course, uses renewable energy in the form of a heat pump for heating and cooling, in combination with a large solar system.

Jürgen Bonin formulated, and continues to develop, his training and project planning documentation during his work as a consultant. Together with his teaching activities at the Oberhausen Chamber of Crafts (HwK), his expertise provides the basis for this handbook on planning heat pump installations.

Jürgen Bonin

The author is also a member of the *Wärmepumpen NRW* and *Geothermie NRW* working groups on heat pumps and geothermal energy in North-Rhine Westphalia. Here, participants regularly give lectures and workshops, and the information gained during these workshops enriches the content of this specialist handbook.

In September 2011, Jürgen Bonin was awarded the energy award at the annual ReneXPO energy trade fair in Augsburg for his invention, the Geo-Protector. The Geo-Protector significantly increases groundwater protection levels when operating brine-water and water-water heat pumps.

Foreword

We are delighted be able to present you with an English edition of this book, after the first German edition sold out so quickly. This reworked edition is a significantly expanded and updated version of its predecessor.

The book is aimed at homeowners and anyone interested in the subject matter in general, as well as heating engineers, planners, architects and urban planners. It is also suitable as a textbook for training and continuing education courses.

There are many types of heat pumps and heat pump systems on the market. The task is to select the right heat pump for the application in question. The choice is large, but the focus must be on selecting the heat pump best suited to the particular application. Not all types of heat pump will be suitable.

Constantly rising energy costs make it increasingly necessary to use renewable energy, especially where this is also becoming a legal requirement. The most elegant form of renewable energy is geothermal energy, because it is available to all of us, right outside our front door and at virtually no cost.

Now the task is to examine the most sensible way of using this geothermal energy, or ambient heat. This is a subject that architects, planners, energy consultants, mechanics and well engineers – in short, everyone whose profession brings them into contact with heat pump systems – need to be familiar with in order to advise their customers accordingly. This book is written to give the user – from planner to heat pump operator – access to the most important and most frequently requested information, as quickly, clearly and simply as possible. Many diagrams and examples illustrate the information provided.

A few ecological observations illustrate the necessity of using renewables. Heat pumps are becoming increasingly important because, when installed with the correct dimensions, they can make a vital contribution to protecting the environment. We should also be aware of the economic reasons for growing heat pump demand. However, badly planned heat pump systems can have precisely the opposite effect.

In writing this book, the author has made every effort to consider all the latest available technologies, as well as current laws and regulations. However, we all recognise the speed of technological advances and the frequency with which legal requirements change, quite apart from the risk of errors, and therefore no guarantee is given for the accuracy or completeness of the information provided. The same applies for the manufacturers' technical details, which should only be taken as examples. When planning heat pump installations, priority must be given to the information and data provided by each manufacturer. The same applies to changes in laws and regulations.

Finally, the author would like to take this opportunity to thank his wife, Heidemarie, and all his partners, customers and colleagues for their many pointers, corrections and tips, as well as the LOHRConsult engineering office and heat pump institute, and other partners and manufacturers, for making their documentation available. The author would also like to thank the Beuth Verlag GmbH for publishing the German edition of this book.

I hope that reading this book gives you plenty of new information and ideas, and I wish you every success in planning your heat pump systems.

Yours,
Jürgen Bonin

1 Introduction

This book provides a detailed examination of everything relating to heat pumps, and takes into account the most recent developments.

We start with a few observations on ecology and economics.

The reader will find in-depth information about heat pump functions and components, including the characteristics of the various refrigerants. The various types of heat pump are described, and their advantages and disadvantages outlined. Other important areas include an examination of performance, coefficient of performance, seasonal performance factor and energy input factor. It is the author's intention to increase awareness of the differences between heat pumps and conventional heating systems. Many hydraulic illustrations of heat pump systems are also provided.

A further, newer aspect is environmental and groundwater protection during heat pump operation. Here, we discuss a variety of solutions.

One reader of the first edition found that the book described water-water heat pumps in a very positive light. This is correct. In contrast to the general trend towards brine-water or air-water heat pumps, more importance should be attached to water-water heat pumps because they achieve the optimum degree of efficiency. The author was recently invited to Gelderland in the Netherlands to view a water-water heat pump system in Arnheim with a total output of 2.3 MW! A range of lectures given on the occasion made it clear that the local inhabitants 'don't want glycol in the ground'! Our neighbours clearly prefer water-water heat pumps as a means of achieving the energy transition. Of course, other countries will have geological and hydrological conditions different to those in the Netherlands, but we can still learn from one another. And now there is an invention that makes 'no glycol in the ground' possible, even for brine-water heat pumps.

The design, planning and selection of hydraulic systems are also extremely important for a heat pump system. As this is where mistakes often occur in practice, it is a subject that will be examined in detail. Planning examples and sample calculations will also be shown.

You will also find helpful hints and explanations on putting heat pump systems into operation.

Chapter 10 discusses whether a heat pump system makes economic sense.

Numerous illustrations and practical examples explore the many issues concerning heat pumps.

2 Observations on ecology and economics

2.1 Environmental impact

In the new, twenty-first century, two trends in particular are leading to a general change in attitude:

1 the striking increases in energy prices; and
2 the consequences of climate change.

Increasingly, housebuilders, landlords and tenants are considering ways of reducing the extra costs associated with housing. It is incredible, but also very natural, that our thoughts turn first to the impact on our wallets.

Sadly, too little attention is still being paid to increasingly striking climate change because it does not yet have a direct impact on our personal finances. But this can change fast – when, or even before, we, or *you*, are directly affected.

Figure 2.1 clearly shows the change in climate.

The future is looking warm!

This will have many recognised and frequently cited consequences:

- melting glaciers;
- rising sea levels;
- polar bears drowning and losing their habitats;
- more species of flora and fauna becoming extinct;
- increasing desertification;
- changing climate; and
- more severe storms and hurricanes, and more.

The advantage of global warming is that less energy will be required for heating, but more electrical power will be needed for cooling, ventilators and air conditioning.

This raises the question: Is this what we want? Of course not. But what can we do about it?

Added to this comes the increasing cost of energy and the threats of oil or gas taps being turned off (i.e. dependence on exporting countries).

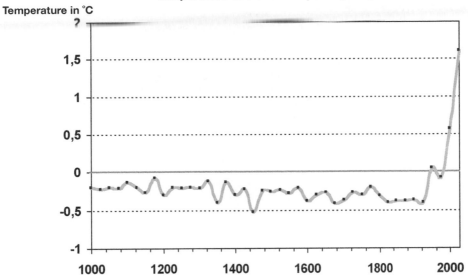

Figure 2.1 Extreme temperature rises at the turn of the century

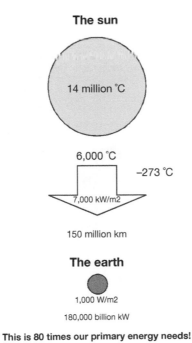

Figure 2.2 The sun as a source of energy

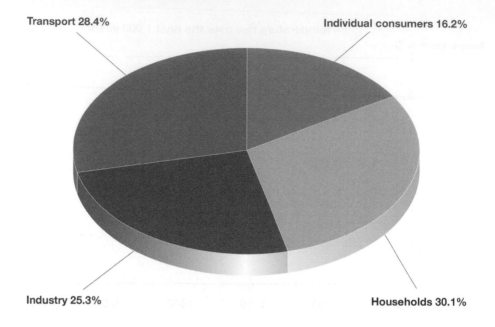

Transport 28.4%

Individual consumers 16.2%

Industry 25.3%

Households 30.1%

Figure 2.3 Energy consumers by share

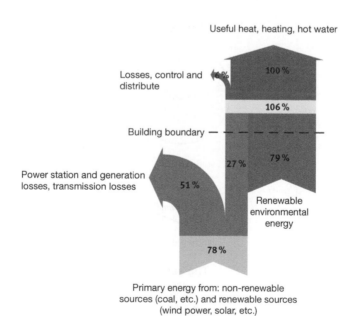

Useful heat, heating, hot water

Losses, control and distribute

6%

100 %

106 %

Building boundary

Power station and generation losses, transmission losses

51 %

27 %

79 %

Renewable environmental energy

78 %

Primary energy from: non-renewable sources (coal, etc.) and renewable sources (wind power, solar, etc.)

Figure 2.4 Energy flow in a heat pump

Is there no alternative? Of course there is.

The only way out of this dilemma is greater use of renewable energy. This includes solar power, without which there would be no geothermal energy, because the earth is warmed by the sun. And the sun provides us with 80 times the energy we actually need (see Figure 2.2).

This energy is available for use in a variety of forms:

- through the direct use of solar energy using solar collectors and solar cells;
- through the immediate use of solar energy in the ground using heat pumps; and
- through the use of wind energy.

Figure 2.3 shows that almost one-third of all energy is consumed by households. The largest portion – 88 per cent – is used for heating.

These observations indicate how necessary it is that we start thinking in ecological terms. Figure 2.4 shows that there is also good financial reason for doing so. Using renewable energy makes both ecological and economic sense.

TIP

Those whose primary energy requirements are too high can use the *green electricity* generated by the major power utilities!

'The streets are paved with gold', by which we mean that geothermal energy is directly available, free of charge and in unlimited quantities, right outside our front doors. You only have to pump it out and use it. Geothermal is one of the ultimate forms of renewable energy and is available to everyone. Unlike all other fuels, there is no need to spend – sometimes considerable – amounts of money and energy transporting it to the end user!

And when you consider that 40 per cent of energy in Europe is used for heating and cooling, then the huge importance of heat pumps becomes clear.

Now we need to rethink energy use in Germany. Although the tendency is increasing, only 20 per cent of new builds are currently equipped with a heat pump. In Austria, in contrast, 50 per cent of all new builds have a heat pump, in Switzerland 85 per cent and in Sweden an incredible 95 per cent. This reduces CO_2 emissions and takes the pressure off consumers' wallets, because:

- like solar energy, geothermal energy is free; and
- reduced CO_2 emissions mean a better climate and act as a brake on climate change and the natural catastrophes it causes, which cost billions in consequential damages.

When God created our world, He bound up CO_2 and other damaging gases in the form of coal, oil and gas, creating a 'good climate' for mankind. And what does mankind then do? Mankind is stupid enough to dig up all these valuable resources out of the ground again, burn them and destroy the climate. It does not have to be like this; indeed, it should not be like this. And it is certainly unnecessary because there are many alternatives – one of which is using geothermal energy by means of heat pumps!

We should also consider that valuable resources, such as coal, gas and especially oil, are finite, and if we carry on as we are now, these will soon be exhausted. And what then?

Heat pumps make both ecological and economic sense. They protect the environment far more than all other fossil fuel-powered methods of heating, and when the heat pump is driven with green electricity, then CO_2 emissions are almost zero.

Heat pumps are the most modern and comfortable heating systems!

For as long as the sun shines, geothermal energy is available to everyone, in unlimited quantities.

The task now is to use heat pumps properly (i.e. in an ecological and economical manner). This requires trained specialists.

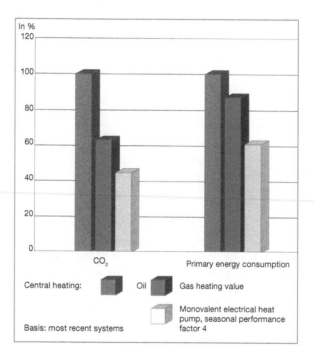

Figure 2.5 Comparison of heating systems

2.2 Economic observations – rising energy costs

Heat pumps not only play an important part in protecting the environment; they are also kind to the operator's pocket. Despite the relatively high acquisition costs, heat pumps quickly pay for themselves. Compared to all other heating systems, heat pumps are the most economical form of heat generation. They also offer the luxury of natural cooling – something no boiler can provide.

Heat pumps need neither oil nor gas!

Why should we continue to remain dependent upon these finite resources? It makes absolutely no sense, especially as *global* demand for these finite, soon unavailable, resources is dramatically increasing.

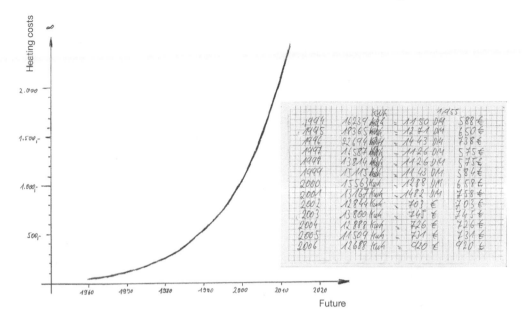

Figure 2.6 Rising heating costs

Nor is the future looking rosy when it comes to increasing energy prices. The past has shown that energy costs double roughly every 10 years (see Figure 2.6).

This development is confirmed by the simple notes taken by any homeowner.

The cause is strongly rising raw material costs for oil and gas (see Figure 2.7).

Figure 2.7 Increasing crude oil prices

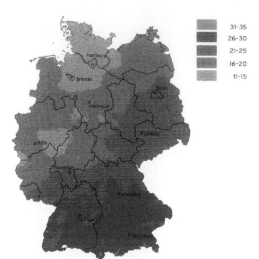

Figure 2.8
Total monthly global radiation in kWh/m² in
January 2004

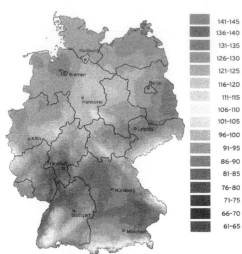

Figure 2.9
Total monthly global radiation in kWh/m² in
April 2004

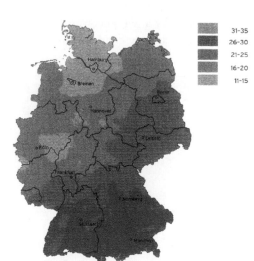

Figure 2.10
Total monthly global radiation in kWh/m² in
July 2004

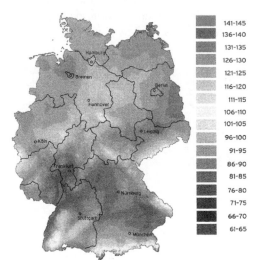

Figure 2.11
Total monthly global radiation in kWh/m² in
October 2004

Who can still afford this?

Based on past experience, you do not need to be a prophet to foresee these price rises continuing into the future. Prices are currently exploding, and no one sees an end to the increasing global demand for energy.

2.3 Where geothermal energy comes from

'Geothermal' derives from the Greek meaning 'earth heat', because the earth is heated by the sun. Without the sun's radiation, earth would be a frozen planet. The sun's radiation provides sufficient heat for any number of heat pumps, as Figures 2.8–2.11 show.

These maps show that there is sufficient energy available to operate heat pumps. For example, even under unfavourable conditions, in the middle of the year at least 150,000 kWh/m²a is available on a plot of ground with a surface area of 200 m² – much more than is required to heat a family home.

3 What is a heat pump and how does it work?

A heat pump is a 'refrigerating machine', a machine usually used for cooling (e.g. a refrigerator). It works by removing heat from an interior space through cooling. This extracted heat must subsequently be released elsewhere. In a refrigerator, this is achieved using a cooler (condenser), which emits the heat into the ambient air.

Heat pumps or refrigerating machines have a long history. The first safe and functional ammonia-refrigerating machine was built by Carl von Linde in 1876. This marked the start of the combined use of cold and heat.

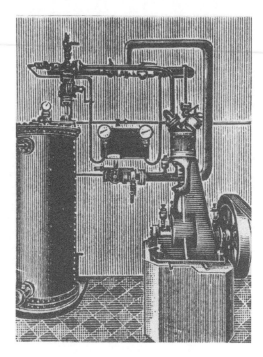

Figure 3.1 1876 – Carl von Linde built the first refrigerating machine

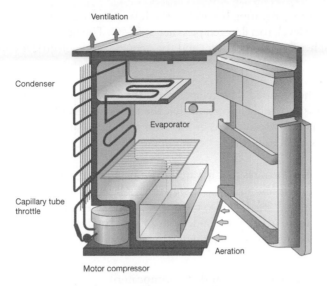

Figure 3.2 The refrigerator – a heat pump refrigerating machine

3.1 Why is a heat pump called a 'heat pump'?

The term 'heat pump' can be explained using the following diagram.

A heat pump 'pumps' heat from a lower temperature level to a higher temperature level.

Figure 3.3 shows a water-water heat pump: it extracts heat from groundwater with a temperature of 10 °C by cooling it to 7 °C. This extracted heat is used, for example, for heating water to a flow temperature of 35 °C. After being used for underfloor heating, for example, the heating water subsequently flows back to the heat pump at 5 °C lower (i.e. it is returned at 30 °C).

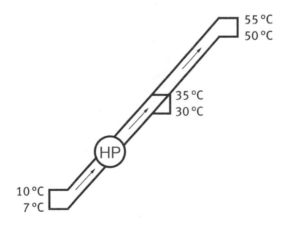

Figure 3.3 Temperature levels in a heat pump

Thus the heat pump has pumped heat from 10 °C to 35 °C. The heat pump has to work in order to reach this temperature level. This requires energy, usually in the form of electricity drawn from the mains supply.

When it comes to generating hot water, the heat pump needs to 'pump' the 'heat' to a higher temperature level, to a flow temperature of 55 °C. The heat pump has to work harder to reach this higher flow temperature, and this requires more electrical energy.

That means:

- The higher the required flow temperature, the harder a heat pump needs to work. Therefore, it is extremely important that the temperature difference between the heat source (e.g. groundwater, brine or air) and thermal heat (e.g. flow temperature for heating) is as small as possible.

This observation demonstrates that a water-water heat pump with a source temperature of around 10 °C will need less auxiliary energy than a brine-water heat pump with a source temperature of around 0 °C. It is also clear that the flow temperature should be as low as possible. This is easy to achieve for underfloor or wall surface heating. It is also clear that it makes little or no sense to use a heat pump to heat radiators with high flow temperatures.

3.2 Structure of a heat pump and its components

A heat pump consists of the following key components:

- compressor with drive motor;
- condenser;
- thermal expansion valve; and
- evaporator.

The diagram below shows the principle structure of a heat pump and its energy flows. This simplified flow diagram also serves as the basis for future explanations.

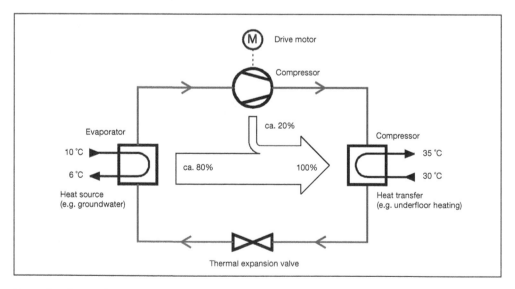

Figure 3.4 Principle structure of a heat pump

3.3 The technical refrigeration cycle and the function of a heat pump

The flow diagram shows the principle structure design of a refrigerating machine and its energy flows. However, in order to ensure the safe operation of a refrigerating machine, several other components are required: a refrigerant collector, dryer, observation glass, and monitoring and protective components (e.g. high and low pressure cut-out switches).

Within the refrigeration cycle is a hermetically sealed gas, the so-called refrigerant. This refrigerant is pumped through the refrigeration circuit by a compressor. This is what we call the Carnot cycle. The individual components have the following functions (see Figure 3.5).

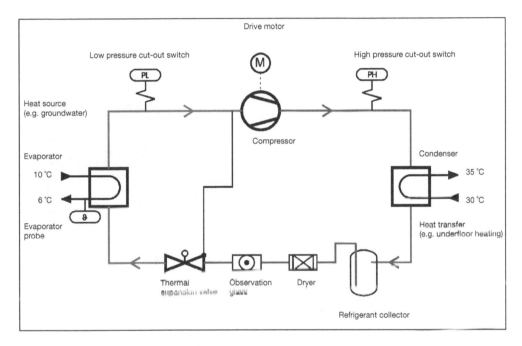

Figure 3.5 The components of a heat pump

1. The compressor with drive motor

The compressor compresses the refrigerant and forms the heart of every heat pump. There are various types of compressor: rotary screw or scroll compressor, or piston compressor. Rotary screw compressors have no oscillating parts and therefore run more smoothly. Consequently, they are quieter than piston compressors.

Every gas heats up when compressed. The greater the compression, the higher the rise in temperature.

This process can be compared to an air pump in which the bottom heats up when pumped. This temperature rise is not the result of friction, but of the gaseous-mixture air being compressed. For a bicycle pump, this is around 0.5–1.5 bar; for a heat pump compressor, around 20 bar, and even more depending on the refrigerant being used. The refrigerant itself accordingly becomes hotter – sufficiently hot for heating and generating hot water.

2. The condenser

The condenser is a heat exchanger, usually a large-scale plate heat exchanger. The gas flows into the condenser where it is cooled by the colder heating water. The gas condenses in the condenser (i.e. it liquefies). The heat emitted through condensation in the condenser is transferred to the heating system.

During this process, the actual temperature changes little, or not at all; instead, we have a change in state, from a gas to a liquid. It is this change of state that releases large quantities of heat energy for heating – far more than if heat were simply extracted.

The refrigerant is now a liquid.

Figure 3.6 Heat exchanger   *Figure 3.7* Thermal expansion valve

If insufficient heat is removed in the condenser, then compression causes an increase in temperature, the so-called discharge temperature of the gas. The result is an increase in pressure on the high pressure side. If a maximum pressure is exceeded, then this leads to a high pressure fault, activating the high pressure cut-out switch. Normally, the control unit then shuts down the heat pump. If insufficient heat is extracted, then the discharge temperature (and consequently also the pressure) can increase dramatically.

If, after the refrigerant has been liquefied, its temperature is lowered further by more heat being extracted, then we talk of the refrigerant being sub-cooled. This results in a little more heat being transferred via the condenser and leads to a minimal increase in the performance of the heat pump.

3. Thermal expansion valve

The thermal expansion valve has a very small opening through which the liquid refrigerant is pressed under high pressure. After passing through the valve, the refrigerant expands. Both the pressure and temperature of the refrigerant fall significantly.

This process is comparable to refilling a cigarette lighter with gas; the gas gushes out, significantly cooling down the lighter. The same thing happens when the gas passes through the expansion valve, but with a far greater difference in pressure (e.g. from 15 bar to 4 bar, dependent upon refrigerant type).

Expansion valves regulate the injection of liquid refrigerants into the evaporator. In turn, this injection is controlled by the overheating of the refrigerant behind the evaporator. Where there is overheating behind the evaporator, the refrigerant not only changes state, but there is also a rise in temperature (overheating). The higher the temperature behind the evaporator (overheating of the refrigerant), the more the expansion valve opens, to inject more refrigerant. The more refrigerant that is fed into the evaporator, the more the temperature behind the evaporator is reduced. If the temperature behind the evaporator becomes too low, then the expansion valve closes, injecting less refrigerant. Thus the role of the expansion valve is to control the flow of refrigerant into the evaporator.

In principle, we differentiate between a thermal expansion valve and an electronic expansion valve. Thermal expansion valves have become the standard valves used in heat pump construction. With thermal expansion valves, the temperature is measured before entering the compressor.

4. The evaporator

The evaporator is also usually a plate heat exchanger. Here, the very cold liquid refrigerant flows into the heat exchanger at temperatures significantly below 0 °C. The refrigerant is heated – and thus evaporates – by the medium that is pumped to the exchanger, either brine (0 °C – brine-water heat pump) or water (10 °C – water-water heat pump), or by the ambient air (air-water heat pump).

During this process, the refrigerant state changes again, from a liquid into a gas – it evaporates. In doing so, the refrigerant absorbs a lot of heat energy, which is later available for heating. Here again, the refrigerant absorbs much more energy by changing state than it would by simply being heated.

If the refrigerant in the evaporator is additionally heated, then we talk of overheating the refrigerant (the opposite of sub-cooling behind the condenser).

In the evaporator, it is important that the refrigerant is evenly distributed within the heat exchanger, as shown in Figures 3.8 and 3.9.

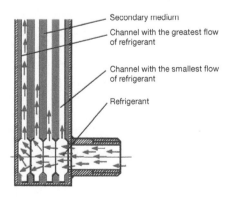

Figure 3.8
Evaporator without refrigerant distributor

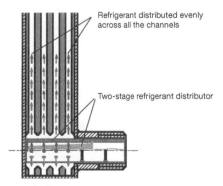

Figure 3.9
Evaporator with two-stage refrigerant distributor

Figure 3.10 Refrigerant collector

5. The refrigerant collector
The refrigerant collector effectively acts as a buffer, holding the liquid refrigerant after it leaves the condenser. It ensures that only liquid refrigerant can enter the expansion valve.

6. The dryer
Despite the most careful evacuation procedures, residual moisture can remain in the refrigeration circuit when a heat pump is manufactured. As this would cause significant impairment to the working of the pump, a dryer is included to remove/bind any residual moisture.

7. The observation glass
The primary role of the observation glass is to help the heating engineer or service technician start up the refrigerator correctly. It allows the technician to see if the refrigerator is working optimally at the various operating points.

The observation glass also allows any residual moisture to be detected, through the colouring of the indicator ring.

8. The low pressure cut-out switch
This switch is an important safety mechanism. Its role is to turn off the refrigerator when the refrigerant pressure on the heat source side is too low. This occurs if the refrigerant is excessively cooled.

There are two principal reasons for this switch to signal a fault:

1   The refrigeration cycle is leaking and refrigerant is escaping. In this case, the low pressure switch must turn off the compressor to stop it from becoming damaged.
2   Cooling on the heat source side is too intense, and therefore the refrigerant is not evaporating sufficiently. Here again, the low pressure cut-out switch turns off the compressor as a protective measure. Should this happen, it is important to examine whether the heat source is sufficiently large. In brine-water heat pumps, the borehole or ground heat exchangers are usually too small. In water-water heat pumps, the water flow is insufficient.

Warning! Here, there is a danger of freezing!

Improvements must be made to the heat source side (brine or water supply, borehole or ground heat exchanger).

9. The high pressure cut-out switch

This switch also has an important safety function. Its role is to turn off the compressor when the refrigerant pressure behind the compressor on the high pressure side is too high, to protect both the compressor and condenser.

This occurs where insufficient heat is extracted from the condenser. This results in too little refrigerant – the so-called hot gas – liquefying, and consequently the pressure on the hot gas side rises. When the refrigerant liquefies sufficiently (condensation), the pressure cannot rise to dangerous levels.

Should a high pressure fault occur, then this does not necessarily lead to an immediate switching off and a high pressure failure. Some types of controller make it possible for a high pressure fault to be recorded only after a time-lapse of a few seconds. As the high pressure switch is a safety switch, it should immediately turn off the heat pump once the high pressure limit is exceeded. Depending upon the type of controller used, it is also possible to activate the turn off mechanism so that the heat pump is only shut down after two or three high pressure faults per day, for example.

The high pressure cut-out switch is a safety mechanism and serves primarily to protect the compressor.

10. Evaporator probe

The controller uses this probe (if present) to record the starting temperature in the evaporator. If this temperature is too low, then this leads to a safety shutdown in order to avoid possible freezing. In a water-water heat pump, the probe will ideally be incorporated into the evaporator so that the safety shutdown can be completed as quickly as possible.

However, this safety probe does not take the place of a flow monitor, which is highly recommended for protecting a water-water heat pump.

11. Controller and probe

The controller, aided by the probe, controls the heat pump and regulates the temperatures specified for the entire heat pump system. For more information, see the following chapter. A few other types of refrigeration cycle are presented and discussed below.

3.3.1 Technical refrigeration cycle in a heat pump with sub-cooling and overheating

The following diagram shows a different method of organising the refrigeration cycle.

In this refrigeration cycle, any refrigerant that has not been fully condensed is sub-cooled in the intermediate heat exchanger to ensure that it condenses. The heat emitted during this process is fed to the refrigerant on the suction side. This, in turn, becomes overheated.

The advantage of this type of heat pump is that there is no need for the refrigerant collector and so less refrigerant can be used. A special intermediate heat exchanger takes its place.

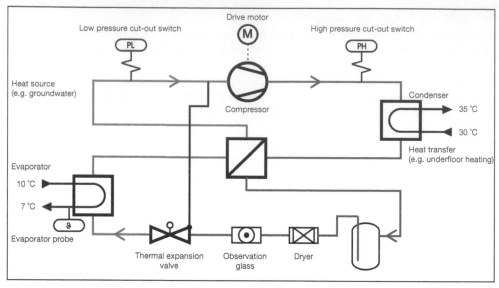

Figure 3.11 Heat pump and its components with intermediate heat exchanger

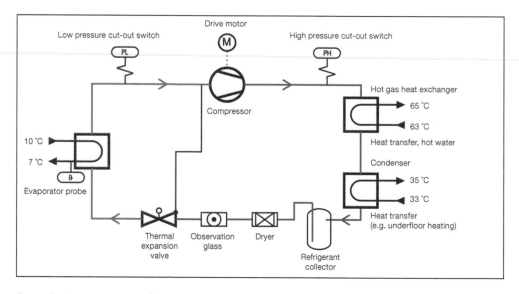

Figure 3.12 Heat pump and its components using the hot gas system

3.3.2 Technical refrigeration cycle for a heat pump with hot gas system

Some heat pumps continue to be equipped with an additional heat exchanger for using hot gas (see Figure 3.12).

The hot gas usually leaves the compressor at very high temperatures – it is hot, and can be used to raise the hot water temperature further. However, it should be noted that when hot gas condenses the refrigerant in the heat exchanger, the pressure of the incoming refrigerant needs to be adequately high. But this lowers the overall efficiency of the heat pump because the compressor's electricity consumption rises accordingly. If this heat exchanger is used 'just' to release some heat in order to raise the hot water temperature, then the quantity of heat given off is proportionally low because large quantities of heat are only emitted as a result of a change in the refrigerant's state. This then corresponds only to the quantity of heat given off by a gas, with the low specific heat capacities typical for gases. Only where there is a change in state will a large quantity of heat be released and transferred. This is the role of the condenser alone.

3.3.3 The basic construction of a heat pump

Figure 3.13 shows the construction of a typical single unit heat pump. The individual components are clearly recognisable.

View of the interior of a *GeoMax®* heat pump:

Electronics ⟶

  1 Main switch

  2 Heat pump controller

Refrigeration technology ⟶

  3 Double thickness sheet steel housing, well sound-insulated

  4 Evaporator

  5 Condenser

  6 Dryer

  7 Chassis

  8 Observation glass

  9 Scroll compressor

  10 Collector

  11 Triple vibration damper

Figure 3.13 Refrigeration technology in a heat pump

The heat pump is particularly quiet due to the double thickness housing with interior perforated steel sheeting and high-quality sound insulation, plus triple vibration damper. In this heat pump, the refrigeration technology is clearly spatially separated from the electronics. All the parts are easily accessible for servicing and maintenance.

The electronics are housed in the upper section of this heat pump, with a continuous terminal strip forming the interface for all electrical connections.

1   Power contactor

2   Motor circuit breaker

3   Terminal strip

4   Cable entry

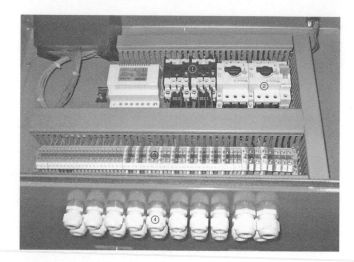

Figure 3.14 Heat pump electronics

The control system also includes other safety mechanisms such as a flow monitor and, if necessary, dry running protection (for water-water heat pumps), as well as flow and pressure monitors (for brine-water heat pumps) and motor circuit breakers, etc.

• Low sound emissions are particularly important in houses with no basement! A well-insulated heat pump is also a sign of quality.

3.4 The controller

Every heat pump must be regulated via a control unit. The controller is responsible for a variety of tasks.

The main task of a controller is to regulate the system in accordance with the weather conditions. This ensures that, depending upon the external temperature, the buffer tank for the underfloor heating is charged only as much as is necessary to ensure a comfortable interior climate. The operator must also be able to make individual settings; they should be able to use the controller to change the heating temperature easily.

The illustration below shows a controller with an open cover. When the cover is closed, the operator can select the type of operation using the upper control knob, and can raise or lower the temperature by +3 °C from the reference value with the lower knob.

ELESTA controller:

1 Selection knob

2 Temperature setting

3 Display with various functions

4 Interface

Figure 3.15 Heat pump controller

Good controllers also offer users the following functions:

- party function;
- holiday programme; and
- cooling function.

There are also various 'service levels' available only to the service technician. Accessible via a code, these enable system-specific values and parameters to be set and altered. However, only a specialist should be allowed access to these levels – someone thoroughly at home with the controller. Incorrect settings can lead to faults or even a breakdown of the entire heat pump system!

For smaller heat pump systems, the setting options outlined above are sufficient.

The controller also has important monitoring functions. In principle, it must monitor the pressure on both the low pressure and high pressure sides. The controller should also have the necessary access points for the motor circuit breaker and other monitoring sensors, such as the flow monitor and pressure monitor. This enables the controller to regulate, control and monitor the entire heat pump system.

However, a heat pump controller should also be able to control more complex systems. A simple example: an older house with underfloor heating on the ground floor and radiators on the upper floors is to be heated using a heat pump. Two buffer tanks are needed to operate the heat pump system properly, one at a lower temperature for the underfloor heating (e.g. 40 °C) and the second at a higher temperature for the radiators on the upper floors (e.g. 50 °C). And of course there is also the hot water tank.

It should also be possible to extend a heat pump controller so that it can regulate larger heat pump systems (e.g. for several mixed and unmixed heating circuits with remote control and swimming pool, etc.). Here, many controllers quickly reach the limits of their capabilities.

The controller should also allow the operator to set the temperatures in individual rooms as required.

3.5 The refrigeration cycle in the p-h diagram

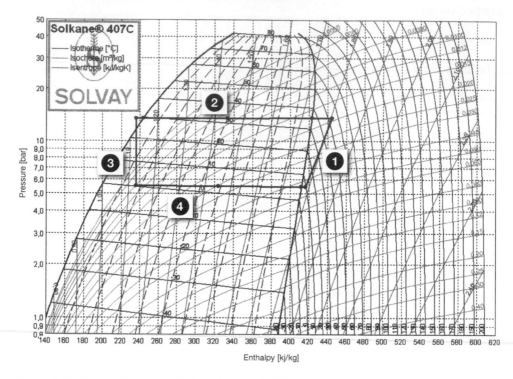

Figure 3.16 Thermodynamic cycle for refrigerant R407c

1 The refrigerant is compressed in the compressor from around 5.5 bar to around 14 bar. This raises its temperature from 5 °C to 35 °C. 2 The refrigerant flows through the condenser and releases energy. 3 The refrigerant flows through the expansion valve and expands, reducing pressure from around 14 bar to 5.5 bar, and cooling from 30 °C to 0 °C. 4 The refrigerant flows through the evaporator and absorbs energy.

1   We start with the compressor: here, the refrigerant is compressed, causing it to heat up strongly. As this change in state runs almost parallel to the isentropic flow (blue lines), it is therefore called an isotropic change of state.

2   The refrigerant liquefies in the condenser. The temperature remains roughly the same, and therefore this is an isothermal change of state.
     The less this process runs parallel to the isothermal curve, the greater the temperature glide. The disadvantage of a temperature glide is described in the following chapter.

3   Here, the refrigerant is forced through the expansion valve and expands. The gas pressure is significantly reduced. Consequently, the refrigerant cools down strongly and still has a large liquid component.

4   Here, the refrigerant flows through the evaporator. Heat energy is transferred to the refrigerant from the heat source and the refrigerant completely evaporates while remaining at almost the same temperature. Therefore, this is also an isothermal change of state.

3.6 The refrigerant

A variety of refrigerants are used in heat pumps, each with advantages and disadvantages. Here, we will examine a few characteristics of three different refrigerants.

Refrigerant R407c

This refrigerant is very often used in heat pumps. It is a mixture of several gases and is described by the curve shown below. The diagram also shows the refrigeration cycle (e.g. as used in a heating system).

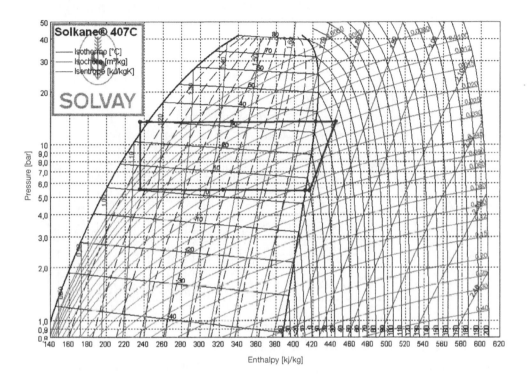

Figure 3.17 Refrigerant R407c in the thermodynamic cycle of a heat pump

Refrigerant R407c has many positive attributes that we will not go into here. Its disadvantage, however, is the 'temperature glide'. The diagram below shows that with refrigerant R407c, the temperature glide can cause the temperature at the entrance to the evaporator to fall very quickly below 0 °C. In Figure 3.18, we see that the lower line of the thermodynamic process crosses the 0 °C line. Below the 0 °C line, there is a danger of freezing!

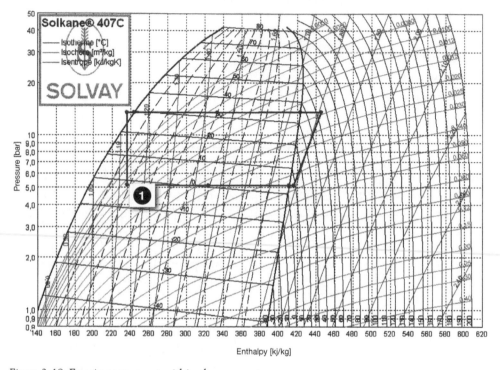

Figure 3.18 Freezing can occur within the evaporator

1 Where the refrigerant dips below the isothermal curve for 0 °C, there is a danger of freezing in the evaporator!

Note on temperature glide:

Temperature glide can be identified where the isothermal curve does not run parallel (horizontal) to the lines of the evaporator. This is completely different for many other refrigerants, for example R134a (see Figure 3.19), where the isothermal curves are horizontal. For this reason, refrigerant R407c can be dangerous if used in water-water heat pumps. Therefore, special safety measures must be considered – for water-water heat pumps, the minimum flow rate of the well water. A suction pressure regulator built into the heat pump can also reduce the risk of freezing.

This is less of a problem for brine-water heat pumps because, thanks to its antifreeze (Antifrogen N or L), the brine can freeze only at very low temperatures. With a correct antifreeze mixture, the heat pump is more likely to cut out due to a low pressure fault.

We can also see that the maximum temperature at 30 bar is limited to around 65 °C. Here, the compressor quickly reaches its performance limit. A high pressure cut-out switch protects the compressor from overloading. Naturally, higher pressures can produce higher temperatures, but in practice this is not expedient when using this refrigerant.

Refrigerant R134a

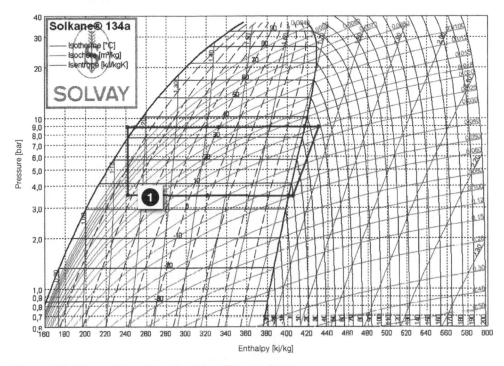

Figure 3.19 Thermodynamic cycle with refrigerant R134a

1 The refrigeration cycle in this diagram shows that the isothermal curves run almost horizontal because the refrigerant has no temperature glide. It is therefore well suited for use in water-water heat pumps.

The diagram also shows that it is easy to achieve higher temperatures with this refrigerant. This would be a great advantage for operating a heat pump for high flow temperatures. However, we should not forget that it is significantly less efficient to operate heat pumps working at high temperatures.

The disadvantage of this refrigerant is the need for very large compressors. This has an impact on the price not only for the compressor, but also for the larger housing this requires.

Consequently, this refrigerant is used only in a few heat pumps.

Refrigerant R410a

This refrigerant is also often used in heat pumps. It has many characteristics that make it suitable for operating a heat pump.

This is another refrigerant in which the isothermal curve is almost horizontal (no temperature glide), making it most suitable for water-water heat pumps. The advantage of this refrigerant is that the compressors are much smaller. The disadvantage is the high operating pressure, somewhat above 40 bar, which requires that the condenser and other components be specially designed for these pressure levels.

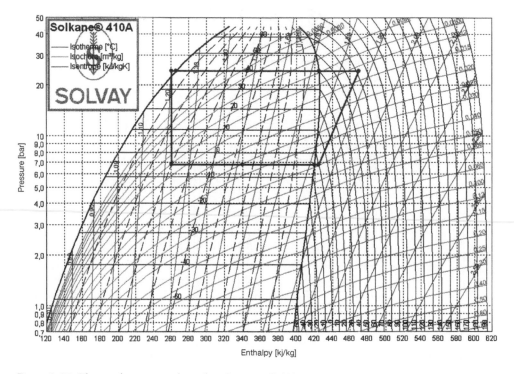

Figure 3.20 Thermodynamic cycle with refrigerant R410a

Another advantage is that the compressors can be small. In turn, the heat exchangers need to be somewhat larger and especially designed to cope with the higher pressures. This has a knock-on effect of increasing the price of the heat exchangers.

PLEASE NOTE

There are also combustible refrigerants (R290), which should no longer be used for safety reasons.

Due to a law passed in Germany in 2006 banning the use of chemicals that damage the ozone layer (Ozone Layer Chemicals Order), only chlorine-free and FCK-free refrigerants may be used. This law regulates the use of chemicals and the management of waste. It is designed to limit the escape of ozone-depleting substances and is thus an important contribution to protecting the ozone layer. National regulations on chemical use must always be adhered to during the repair and servicing of refrigeration machines.

3.7 Water-water heat pumps

A water-water heat pump extracts heat from water, usually well water from a bore well.

1   Buffer tank

2   Hot water tank

3   Heat pump

4   Intake well

5   Discharge well

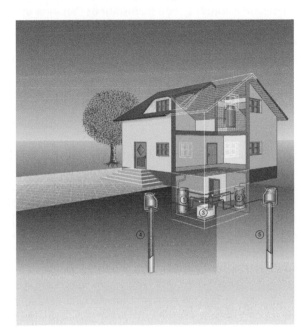

Figure 3.21 Water-water heat pump system

This form of extracting energy is called hydrothermal energy generation. The hydrothermal energy generation process extracts heat energy from water pumped up from the intake well. This causes the water to cool down. We can generally assume that well water will usually have a temperature of around 10 °C, and – as the heat pump extracts heat energy – the water normally cools by around 3 K. It is then returned to the ground via the discharge well. The major advantage of this system is that the heat pump is continually fed with water at a temperature of around 10 °C.

This illustration demonstrates that the intake and discharge wells must be sufficiently far apart in order to prevent a 'thermal short circuit'. An underwater pump built into the well usually pumps the well water up to the heat pump.

The illustration also shows all the key elements in a water-water heat pump system:

- intake well;
- discharge well;
- heat pump;
- buffer tank; and
- hot water tank.

When considering a water-water heat pump, it is important to consider the quality of the well water.

Where the iron or manganese content of the groundwater is too high, there is a danger of the evaporator and discharge well becoming encrusted (clogged). If the evaporator becomes clogged, there is a danger that it may freeze, causing the heat pump to shut down. If the discharge well is clogged, then it cannot absorb more water and it floods. The encrustation is caused by insoluble hydrated oxides precipitating out of water with a high iron and manganese content. The incrustations can clog up wells and components in the heat pump system.

Where the water is corrosive, often a nickel-brazed heat exchanger is recommended. A borehole heat exchanger is also advisable for use on sites with very corrosive water.

Furthermore, if using a water-water heat pump, an adequate supply of water must be guaranteed. A flow monitor is advisable for safety reasons, set to the minimum required flow. The safety of the system can be increased by adding dry running protection. Where the supply of water to the heat pump is insufficient, then the evaporator is in danger of freezing.

The 'cold side' of a water-water heat pump system is excellently suited for providing a building with 'free cooling'. This means that the heat pump compressor does *not* need to work. The 'cold side' – the cool groundwater – can be used for cooling via a heat exchanger. Cooling can also be achieved using a reversible heat pump. Here, the heat pump extracts heat from the house and directs it back into the ground. This requires the heat pump compressor to operate, and this needs additional power – degrading the overall energy balance.

The advantages of a water-water heat pump:

- The high costs associated with a borehole heat exchanger can be avoided.
- The heat pump system is significantly more efficient – around 20 per cent more efficient than a brine-water heat pump.
- It is ideal for heating and cooling.
- It is optimal for free cooling.
- Extra water is available free of charge for irrigation, etc.

The disadvantage of a water-water heat pump:

- Water containing iron and manganese, or very corrosive water, is unsuitable for use in a water-water heat pump. Subterranean (underground) water preparation methods can be used where water contains iron and manganese.

3.7.1 Water-water heat pump with no system separation

Figure 3.22 shows a water-water heat pump with no system separation.

In this heat pump system, well water flows directly through the evaporator. This ensures an optimal efficiency factor because of the high source temperature of the well water (around 10 °C). The disadvantages lie in the increased risk involved:

1   Greater danger of freezing due to insufficient flow through the evaporator.
2   Corrosive water leads to an increased risk of the evaporator corroding. Normally a brazed heat exchanger should be used here. However, this can be problematic for

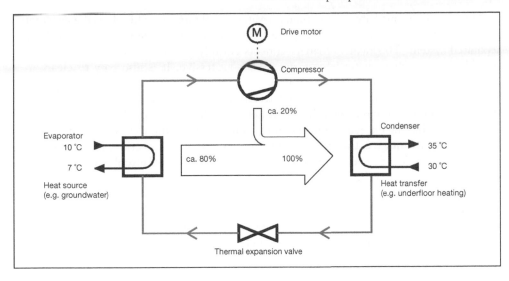

Figure 3.22 Water-water heat pump with no system separation

larger heat pump manufacturers if this involves a deviation from mass production. Smaller heat pump manufacturers can often be more flexible in this respect.
3 Should there be a leakage (e.g. if the evaporator becomes defective), then this poses a danger to the environment if refrigerant and/or oil are able to escape into the groundwater.

These risks can be almost excluded by using a heat pump with system separation.

3.7.2 Water-water heat pump with system separation

Figure 3.23 shows a water-water heat pump with system separation.

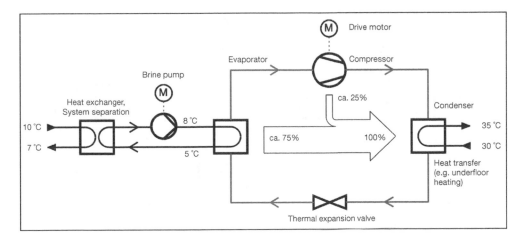

Figure 3.23 Water-water heat pump with system separation

The closed intermediate cycle contains a water-glycol mixture. This offers the evaporator protection from freezing. A bolted plate heat exchanger is often recommended for a separated system. This heat exchanger consists entirely of stainless steel and is therefore suitable for use where water is corrosive and/or aggressive. It is also easy to clean, as a bolted plate heat exchanger can be easily taken apart, cleaned and reassembled. The danger of the heat exchanger freezing in a separated system is very low, because before this happens the heat pump would have been shut down, either as the result of a low pressure fault, or of refrigerant having exceeded a temperature limit upon leaving the evaporator.

A further advantage is the greater level of environmental protection. If there is a leak, then a patented Geo-Protector can be used to suck up any escaped glycol or oil, and remove it from the discharge well.

The disadvantage is the reduction in the heat pump system's overall level of operating efficiency. On the one hand, the efficiency of the heat pump is reduced by around 4 per cent as a result of the lower entry temperature in the evaporator, and, on the other, additional energy is required to run the second circulation pump.

3.8 Brine-water heat pump

This is another form of geothermal heat generation. Here, the brine-water heat pump draws its heat not from well water, but rather from the ground, via a closed brine circuit. The brine circuit contains a glycol-water mixture to prevent it from freezing in the heat pump evaporator. It has a concentration of around 30 per cent to 35 per cent. The brine circuit has the effect of cooling the earth immediately around it, and therefore the temperature of the heat source for the heat pump is accordingly lower than for a water-water heat pump. Thus a brine-water heat pump needs to overcome a greater temperature difference, and requires more energy (i.e. electricity) to do so. The 'heat' extracted from the ground is fed to the heat pump via a closed brine circuit.

Diagram of a brine-water heat pump:

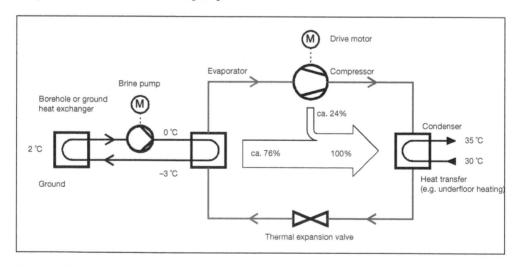

Figure 3.24 Diagram of a brine-water heat pump

Brine-water heat pumps are divided into those with borehole heat exchangers and those with ground heat exchangers. Both are usually manufactured from particularly durable PE pipes, making them almost an investment for life.

IMPORTANT

- When using brine-water heat pumps, attention must be paid to any local installation restrictions, especially close to drinking water extraction areas.
- A pressure monitor must be installed to ensure that the brine circuit is not losing any brine. Where there is an undue fall in pressure, the monitor prevents the heat pump from being turned on.

3.8.1 A brine-water heat pump with borehole heat exchangers

Borehole heat exchangers are inserted vertically into the ground. When the heat pump operates, the borehole heat exchangers extract heat from the surrounding ground. This leads to a cooling in the immediate soil area. Consequently, after a period of operation, this significantly cools down the surrounding ground temperature (steady-state temperature), more so than with a water-water heat pump. The normal operating temperature is 0 °C.

1   Hot water tank

2   Buffer tank

3   Heat pump

4   Borehole heat exchangers

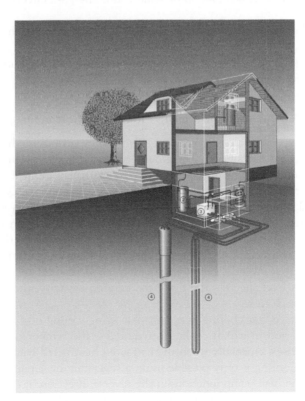

Figure 3.25 Brine-water heat pump system with borehole heat exchangers

The borehole heat exchangers were developed to prevent wells from clogging or suffering corrosion as the result of aggressive water. Thanks to system separation (water/brine), neither is possible. The lower temperature of the brine significantly reduces the overall efficiency of brine-water heat pumps compared to their water-water equivalents.

When planning the use of borehole heat exchangers, attention must be paid both to the cooling capacity, which will be extracted from the ground, and to the local geology, namely the specific abstraction capacity. The specific abstraction capacity is the cooling capacity per drilled metre (W/m). In addition, boreholes should be no closer than a set minimum distance apart (usually at least 5 metres). Deep borehole heat exchangers cannot always be bored completely vertically, and thus the distances between them will vary at greater depths.

As the borehole heat exchangers reach down deep into the 'cool' earth, a brine-water heat pump system with borehole heat exchangers is well suited for the free cooling of buildings.

The advantages of a brine-water heat pump with borehole heat exchangers:

- The danger of clogging and freezing is eliminated, therefore operational reliability is high.
- Good for both heating and cooling.
- The heat source has an extremely long lifespan, effectively an investment for life.

The disadvantages of a brine-water heat pump with borehole heat exchangers:

- Significantly less efficient than a water-water heat pump.
- Very high installation costs.

3.8.2 A brine-water heat pump with ground heat exchangers

A less expensive alternative, especially for those installing their own heat pump systems, is ground heat exchangers. The ground heat exchangers are installed horizontally, beneath the ground. Just as with borehole heat exchangers, the heat exchangers extract heat from the surrounding soil. However, as the heat is being removed very close to the surface of the ground, the ground heat exchangers are subject to marked variations in temperature.

When planning the use of ground heat exchangers, attention must be paid both to the cooling capacity to be extracted from the ground, and to the local geology, namely the specific abstraction capacity. The specific abstraction capacity is the cooling capacity per surface area (W/m^2). Before installation, thought must be given to the distance between the heat exchangers – the lower the cooling capacity, the greater the distance between them needs to be.

Effective 'free cooling' is not possible using ground heat exchangers because in summer – the period in which cooling is required – the temperature of the ground close to the surface rises, especially when the extracted heat is fed back into this area during cooling mode. Cooling is only possible using a reversible heat pump (see reversible heat pump). Here, the heat pump functions in the reverse direction: it extracts heat from the building, transferring it via the ground heat exchangers back into the earth. This requires the use of the compressor. Consequently, the operating costs are higher.

1 Hot water tank

2 Buffer tank

3 Heat pump

4 Ground heat
   exchangers

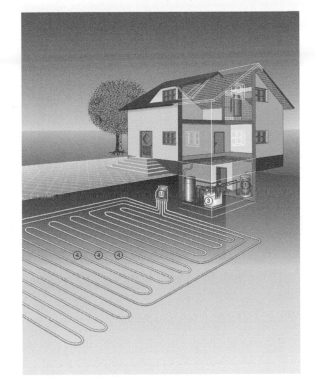

Figure 3.26 Brine-water heat pump system with ground heat exchangers

The advantages of a brine-water heat pump with ground heat exchangers:

• No expensive borehole heat exchangers.
• Good level of efficiency for heating.
• Very high operational reliability.
• Extremely long lifespan – effectively an investment for life.

The disadvantages of a brine-water heat pump with ground heat exchangers:

• Significantly less efficient than a water-water heat pump.
• No free cooling.

3.9 Heat pump with direct evaporator

A heat pump system with direct evaporator using borehole heat exchangers or ground heat exchangers is constructed as follows:

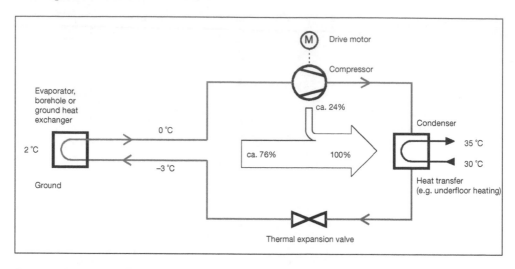

Figure 3.27 Diagram of a brine-water heat pump with direct evaporator

In a heat pump system with direct evaporator, the evaporator is located in the ground, not in the heat pump itself. The evaporator can be composed of the borehole heat exchangers or ground heat exchangers. Consequently, the process of evaporation no longer takes place in the evaporator unit in the heat pump (as in a conventional brine-water heat pump), but rather directly in the ground. This means that having passed through the expansion valve, the refrigerant is fed directly into the heat source – the ground – at a temperature of around −3 °C. Thermal contact with the ground via the borehole heat exchangers or ground heat exchangers heats the refrigerant, causing it to evaporate within the borehole or ground heat exchangers.

The advantage of a direct evaporator is its greater efficiency, because this system removes the transfer losses inherent in using a heat exchanger located within the heat pump. This system uses one less heat exchanger, because the borehole or ground heat exchangers are used as evaporators. Furthermore, this system has no need of a brine circulation pump, nor the power needed to drive it (although the impact here is limited).

The disadvantage is that the very large refrigeration circuit requires large quantities of expensive refrigerant. When installing the borehole or ground heat exchangers, it is vital that they are 100 per cent watertight, to avoid the expense of leaking refrigerant, and therefore borehole and ground heat exchangers being used with direct evaporators require lined metal pipes.

A further disadvantage is that free cooling is not possible without the heat pump operating in reverse.

The advantages of a brine-water heat pump with direct evaporator:

- Higher level of efficiency than a brine-water heat pump.
- High levels of operational reliability.
- Long lifespan – effectively an investment for life.

The disadvantages of a brine-water heat pump with direct evaporator:

- Very high material costs for the borehole/ground heat exchangers and for the refrigerant.
- Less efficient than a water-water heat pump.
- No free cooling.

3.10 Air-water heat pump

The ambient air serves as the heat source for an air-water heat pump.

1 Intake air duct

2 Exhaust air duct

3 Hot water tank

4 Buffer tank

5 Heat pump

6 Evaporator

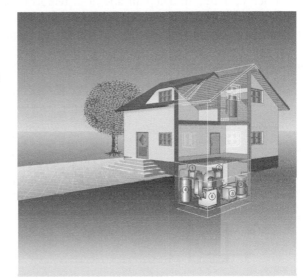

Figure 3.28 Air-water heat pump system

This form of energy extraction is called aerothermal energy extraction. During aerothermal energy extraction, an air-water heat pump extracts heat from the ambient air. Although technically more complicated, an air-water heat pump system is better value (i.e. cheaper) overall. The evaporator and the heat pump are shown separately in the figure above. The large evaporator, with a large, additional ventilator, makes it more technically complicated.

An air-water heat pump extracts heat from the ambient air. As the specific heat capacity of air is very low, relatively large volumes of air need to be circulated. Therefore, it is important that the air ducts on compact systems are of sufficient dimensions. Large ventilators are needed to move the required volumes of air, and these require correspondingly larger amounts of operating power.

The disadvantage of air-water heat pumps is that the heat source – air – is coldest at times of greatest heat demand, and therefore they run at low levels of operational efficiency during these periods. In contrast, they are somewhat more efficient than brine-water heat pumps in spring and autumn. Added to this, the evaporator can quickly freeze, especially on cold, wet days, and needs to be thawed out again. This also requires additional energy.

Air-water heat pumps are not often operated using a single power source, making a secondary source necessary (e.g. an electric heating element or additional oil or gas heating). This reduces the overall efficiency and economy of the system considerably.

All in all, air-water heat pumps are less efficient, especially in cold and damp climates. Air-water heat pumps are divided into:

- compact units; and
- split units.

As air-water heat pumps extract heat from the ambient air, the evaporator can freeze, especially when operating in damp air and/or at temperatures near freezing. The damper the air, the more frequent the freezing, meaning that the evaporator must be regularly thawed out. This costs additional energy and reduces the pump's efficiency correspondingly.

Schematic diagram of an air-water heat pump:

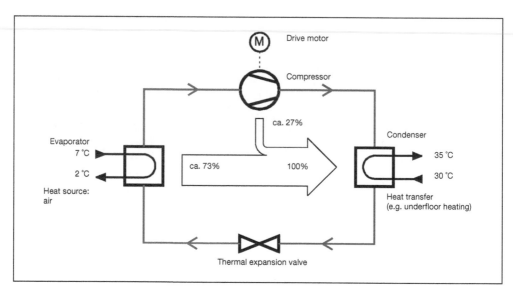

Figure 3.29 Diagram of an air-water heat pump

3.10.1 Compact units

Compact units are single units (i.e. heat pump and evaporator contained within a single housing). Compact units are either for installation indoors or for outdoors. Due to the low specific heat capacity of air, relatively large volumes of air need to be circulated. For compact units installed indoors, the air is fed to, and extracted from, the heat pump via air ducts. These are not required for compact units installed outdoors. The advantage of compact units is their small footprint.

Figure 3.30 shows a compact air-water heat pump for outdoor installation.

Figure 3.30 Compact air-water heat pump for outdoor installation

An example of a compact unit:

1 Outdoor air-
   water heat
   pump

2 Buffer tank

3 Hot water tank

Figure 3.31 An air-water heat pump system

Advantages of an air-water heat pump as a compact unit:

- No costs for wells, borehole or ground heat exchangers.
- Small footprint.

Disadvantages of an air-water heat pump as a compact unit:

- Lower overall efficiency.
- Large air ducts needed for indoor installation.
- When installed outdoors, the entire heat pump technology is exposed to the weather.
- When installed outdoors, the heating water must be frost-proof (e.g. a glycol mixture).
- Monovalent operation often not possible.

3.10.2 *Split units*

Split units are air heat pumps that effectively consist of two parts – the actual heat pump, and the separate evaporator with ventilator. Usually the evaporator and ventilator are installed outdoors to avoid the need for air ducts.

Advantages of an air-water heat pump as a split unit:

- No costs for wells, borehole or ground heat exchangers.
- No large air ducts required.

Disadvantages of an air-water heat pump as a split unit:

- Lower overall efficiency.
- Cooling only possible with reversible heat pumps.
- Monovalent operation often not possible.
- Can be noisy during defrosting.

3.11 Boiler heat pumps

As the name suggests, boiler heat pumps consist of a hot water boiler and a smaller heat pump, often sitting directly above the boiler. These heat pumps are commonly installed in basement boiler rooms and use the waste heat from the heating system to heat water for domestic use.

Boiler heat pumps are ideal methods of using the waste heat from heating systems, especially for existing building stock and houses with high flow temperatures.

Depending upon their make, these boiler heat pumps can also be fitted with air ducts in order to use the ambient air as a heat source.

There are also boiler heat pumps equipped with one or two additional tubular heat exchangers in the tank. The upper heat exchanger connects to an additional boiler, and the lower heat exchanger to a solar system.

3.12 Boiler heat pumps with heat recovery from the exhaust air

These boiler heat pumps are often integrated into ventilation systems. They use the heat from the exhaust air as a heat source for hot water generation.

The disadvantage of this system is that the same quantity of cold air streams into the house as the quantity of hot exhaust air. Where a waste heat recovery exchanger is used, the exhaust air is correspondingly cold and less suitable as a heat source. Where there is no waste heat recovery exchanger, then the cold air streaming in can lead to drafts.

These boiler heat pumps are often an expensive investment and therefore have a relatively long amortisation period.

3.13 Cooling with a heat pump

Every heat pump has a feature that separates it from other boilers or heating units: it has both a cold and a hot side. The hot side is used for heating, and the cold side can be used for cooling.

As more and more houses are insulated, cooling becomes more effective. Insulation protects both against excessive heat and excessive cold.

Cooling ceilings are practical where cooling needs to be particularly effective. A combination of underfloor heating and a cooling ceiling is ideal. Both can be used to heat or to cool. As these are surfaces, both predominantly heat or cool via radiation. Heat rises, making underfloor heating most effective as a means of heating, and cold sinks, so that cooling is most effective using a cooling ceiling.

Combining a heat pump system with controlled aeration and ventilation of the living space using waste heat recovery can increase living comfort. A ventilation system makes cooling the rooms through aeration more effective. The effectiveness can be increased further using a geothermal ground heat exchanger.

Cooling is becoming increasingly important, particularly for commercial use (e.g. in office buildings and hotels, etc.). Cooling is also important for large heat pump systems and heat pumps in residential neighbourhoods, because the process of cooling with geothermal heat pumps returns heat to the ground, 'regenerating' it. See: Heat pumps in residential areas and large heat pump systems.

There are two methods of cooling:

1   The cold side is used directly for cooling, without the heat pump being run. This is what we call 'free cooling'.
2   The heat pump is run in reverse. We refer to this as 'active cooling'. This is the same as reversible refrigeration.

3.13.1 Free cooling

For free or indirect cooling, the cold side is used directly for cooling.

The advantage of this form of cooling is that the heat pump and heat pump compressor are not turned on, and therefore no power is used in cooling. This is therefore the most ecologically and economically advantageous solution.

In the brine-water heat pump shown in Figure 3.32, heat is transferred from the house to the ground during cooling. This subsequently raises the ground temperature; when in constant operation, the ground temperature, and therefore also the brine temperature, rise to above 10 °C. This reduces the effectiveness of the cooling. Even so, cooling via borehole heat exchangers remains possible, although cooling with ground heat exchangers is no longer so efficient.

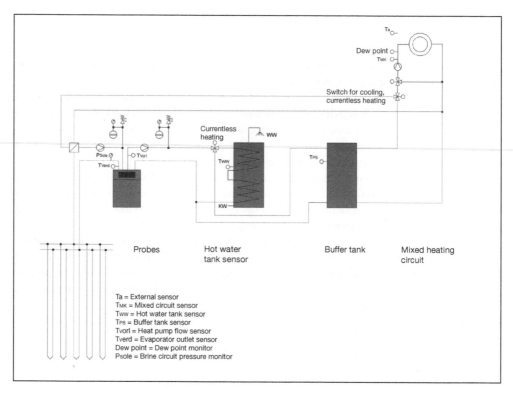

Figure 3.32 Brine-water heat pump system with free cooling – with cooling for the underfloor heating circuit

Free cooling is more effective with a water-water heat pump as here water with a temperature of around 10 °C is constantly available. This form of cooling is ideal for smaller and larger buildings. When used together with underfloor heating, we talk of 'activating the building core', because the core of the building structure is actively used for both heating and cooling.

This offers particular advantages for commercial buildings in which some areas require intensive cooling while others need heating.

3.13.2 Reversible heat pumps for active cooling

Another possibility is 'active cooling', also known as direct cooling. Here, the heat pump is operated in reverse. Heat pumps that can heat or cool according to atmospheric conditions have a four-way switching valve that allows the heat pump to be switched from heating to cooling, and vice versa.

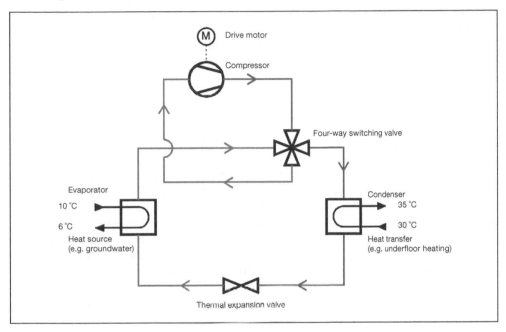

Figure 3.33 Reversible heat pump during heating

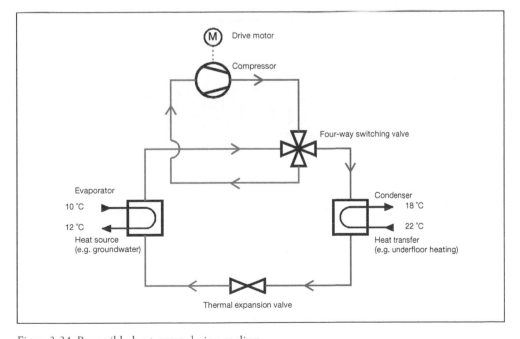

Figure 3.34 Reversible heat pump during cooling

This form of cooling is highly effective.

However, care must be taken that it does not lead to condensation within the building, which can encourage mould. Mould is dangerous to health! Therefore, condensation sensors are recommended in order to prevent condensation. These, in turn, have the disadvantage that, once activated, they permit cooling to restart again only after a very long time (for technical reasons, sometimes no further cooling is allowed). Therefore, the controller must have a cooling function allowing the flow temperatures in the heating circuit to be limited (e.g. ELESTA).

The disadvantage is that the compressor runs during reversible cooling, unavoidably resulting in higher operating costs. Here again, there are advantages when used in commercial buildings in which some areas have high cooling requirements while others need heat.

Being able to cool with a heat pump is an advantage that no other boiler or heating unit offers!

3.13.3 Control during cooling

If the controller in a heat pump is intended to take over the role of cooling in summer, then several preconditions must be met. First, the controller must be capable of regulating the cooling operation. Not every controller can. Cooling generally takes place when atmospheric conditions dictate. This requires a mixed heating circuit, at least for free cooling. Only then can the heat pump controller regulate the flow temperature in accordance with weather conditions. Again, the flow temperature should be kept to a minimum to prevent condensation forming within the building. Naturally, this will depend

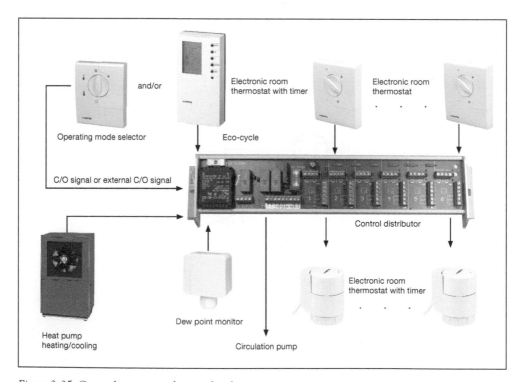

Figure 3.35 Control system cooling with a heat pump

upon the physical characteristics of the building: a house with a greater retention capacity can generally absorb and store more moisture, thereby lowering the dew point. Therefore, in such a building, the minimum permitted temperature can be programmed a little lower than in a building with a lower retention capacity.

Furthermore, so-called cooling ceiling controllers are needed for the individual room controllers. For heating, they open the control valve for the heating circuit when it becomes cooler. The controller for cooling functions in exactly the opposite manner: for cooling, they open the control valve for the heating circuit when it becomes warmer.

This is only possible where the heat pump controller and the cooling ceiling controller can communicate via a simple two-core cable, ensuring that all the controllers act in unison, either to heat or to cool. If, for example, the heat pump controller tries to cool but the individual room controllers try to heat, this can result in individual room controllers determining that it is cooler in the rooms and thus opening the control valves further. This in turn leads to the rooms being too strongly cooled, making controlled cooling impossible. There is then the danger of condensation.

In this system, all the controllers communicate with one another to avoid misunderstandings.

3.14 Gas-driven heat pumps

As well as electricity or gas, heat pumps can also be driven with other fuels including oil. Here we differentiate between gas-powered heat pumps and absorption heat pumps: in the former, the compressor is driven by a gas engine, whereas absorption heat pumps rely on the absorption effect.

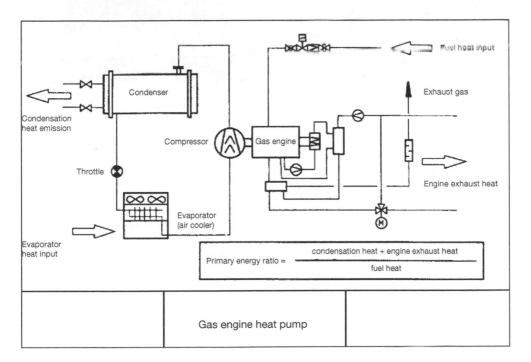

Figure 3.36 Schematic diagram of a gas-driven heat pump

3.14.1 Gas engine heat pumps

A gas-driven heat pump has a gas or petrol engine rather than an electric motor. The exhaust heat from the engine can be completely utilised. These heat pumps have very low energy costs because the losses are minimal and most of the heat energy is drawn from the environment.

However, due to the comparative expense of maintaining the engine, heat pump systems such as this are only suitable for large installations.

Figure 3.37 Gas motor with screw compressor for heat output of 600 kW

3.14.2 Absorption heat pumps

An absorption heat pump is also a refrigerating machine used to generate heat, similar to a heat pump. Instead of a compressor, an absorption heat pump has a so-called 'thermal compressor'. A thermal compressor uses a heating medium such as natural gas or oil, whereas a mechanical compressor requires electricity. Absorption heat pumps are generally driven by natural gas, a technique known from camping or on smaller yachts.

Function of an absorption heat pump:
As in the heat pumps described above, an absorption heat pump primarily consists of a compressor – here, a thermal compressor – a condenser, an expansion valve and the evaporator (see Figure 3.38).

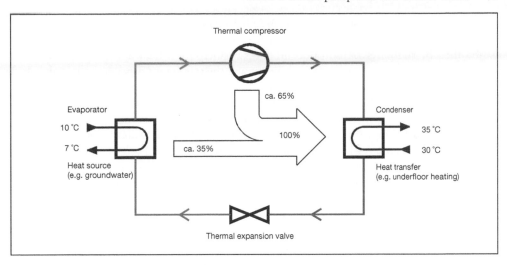

Figure 3.38 Energy flow in an absorption heat pump

The thermal compressor consists of an absorber, a combustion chamber, an expeller and a solvent pump. Ammonia frequently serves as the refrigerant (R717), although lithium bromide may also be used.

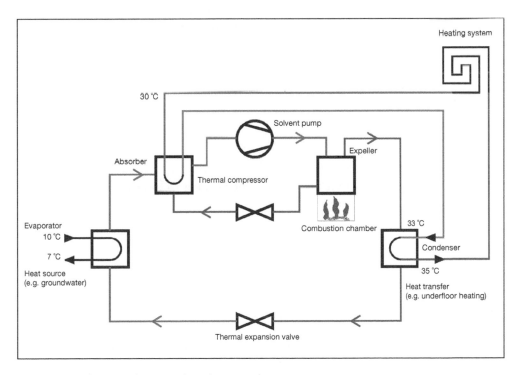

Figure 3.39 Schematic diagram of an absorption heat pump

The refrigerant-poor absorption medium (ammonia water or lithium bromide water) and the refrigerant (ammonia or lithium bromide) meet in the absorber. Here, the refrigerant is absorbed by the water. Absorption produces condensation heat and solvent heat. The refrigerant-poor absorption medium condenses and the refrigerant dissolves as it is absorbed by the refrigerant-poor absorption medium. The heat this produces is transferred to the heating water in the absorber. The refrigerant-enriched solution is then pumped to a higher pressure (e.g. 10 bar) by a small solvent pump. This does not change its physical state. After the solvent pump, the ammonia water or lithium bromide water solution flows into the expeller. Here, the refrigerant (ammonia or lithium bromide) is 'boiled out' at around 170 °C before returning as a gas to the condenser. A fuel, usually gas, is used for boiling, releasing the refrigerant, now in gas form. When it passes on to the condenser (as described in the heat pump above), the condensation heat is transferred to the heating system. The refrigerant expands again after passing through the expansion valve, and can now reabsorb ambient heat in the evaporator. Finally, the refrigerant flows back into the absorber where it is absorbed by the refrigerant-poor absorption agent.

Due to ammonia's toxic nature, special safety measures are required during operation (and particularly during servicing) in the utility room in which the heat pump is installed. Attention should be paid to the German accident prevention regulations UVV VBG 20, DIN 8901, DIN 8875-1 to 8 and EN 378.

As with the absorber fridges used when camping, absorption heat pumps have almost no moving parts. As a result, they have a long lifespan. In absorption heat pumps, the ratio of useful heat output to absorbed burner capacity of the expeller corresponds to the primary energy ratio ξ. Similarly to gas boilers, exhaust condensation is also possible with absorption heat pumps. The primary energy ratios achieved in this way are between 1.5 and 1.7.

$$\xi = P_H / P_B$$

where ξ = primary energy ratio

P_H = heating power

P_B = burner capacity

The ecological advantage here is the very low energy input factor, which is calculated as:

$$e_p = \xi / fp$$

where e_p = energy input factor

ξ = primary energy ratio

fp = energy factor or primary energy factor

As in other heat pumps, the heat source can be groundwater, brine via borehole heat exchangers or ground heat exchangers, or air.

3.14.3 Comparison of both gas-driven heat pumps

Gas-driven heat pumps have a major ecological advantage in that the energy of the energy source (gas or oil) can be almost 100 per cent utilised on site. This is not the case for electrical heat pumps because the process of energy generation in power stations is relatively inefficient. A good, modern power station has an efficiency level of only up to around 40 per cent.

The exhaust heat from a gas-driven heat pump can be used to cool the engine (e.g. by using it to generate hot water). The heat generated by the heat pump can then be optimally used for heating, etc. These heat pumps therefore operate both in an ecologically and economically beneficial manner.

The only disadvantage, however, is the high cost of servicing the combustion engine.

Gas-driven heat pumps are ideal for use in larger heat pump systems, especially where the exhaust heat from the motor can be completely utilised (e.g. to heat swimming pools, etc.).

Absorption heat pumps generally have a level of efficiency – expressed as the primary energy ratio – of only 1.4 to around 1.5. This represents an efficiency of around 65 per cent, maximum 70 per cent. In contrast, when the energy input factor is examined, it is clear that an absorption heat pump offers no additional ecological or economic benefits compared to an electrical heat pump. The only advantage is the long lifespan of a gas absorption heat pump, the result of it having no, or only a few, wearing parts. However, because of their low efficiency levels, these heat pumps are not yet recommended for heating buildings at their current stage of technological development.

3.14.4 Zeolite gas adsorption heat pump

A new heat pump technique was presented in 2009, one that works with zeolite, gas and solar power. The pioneer in this field is the company Vaillant: it has successfully completed field-testing and has been selling this heat pump technology since April 2010. Zeolites have been used as water softeners in detergents since the 1980s, in catalytic converters, and in gas treatment.

The zeolite heat pump is an adsorption heat pump that uses water or steam as the refrigerant. At the heart of this heat pump is a zeolite core. Figure 3.40 shows the zeolite, an ecologically friendly, non-toxic and incombustible ceramic material. Zeolite is able to adsorb (take up) water and release heat in doing so. When subjected to heat, the water is then desorbed (released).

Figure 3.41 shows the complete internal construction of a zeolite module. The cluster of pipes at the bottom is a pipe heat exchanger, which condenses or evaporates the refrigerant water depending upon operating mode. The core of the heat pump is the zeolite module with internal evaporator and condenser.

Another key component is the gas boiler, which is needed for the desorption process: desorption requires temperatures of 120 °C. The heat source here is a solar power unit, and very low flow temperatures of only 3 °C are sufficient for the adsorption phase.

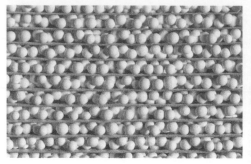

Figure 3.40 Zeolite

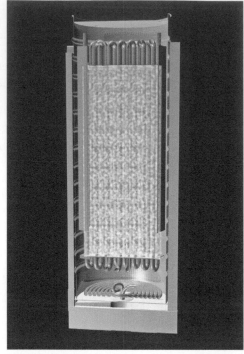

Figure 3.41 Zeolite module

The principal function of a zeolite heat pump is described below: the zeolite module operates with a vacuum of 5 mbar to 200 mbar. This reduces the evaporation and condensation temperature of the refrigerant water.

In principle, there are two operating phases: adsorption and desorption.

The adsorption phase:
During the adsorption phase, solar heat evaporates the water in the lower section of the zeolite module. This does not require a large solar power system: only three flat plate collectors with a total surface area of around 7 m^2 are required. Flow temperatures of only 3 °C are sufficient to evaporate the water in the vacuum of the zeolite module. The pressure in the zeolite module rises as steam forms. The steam is adsorbed by the zeolite and the zeolite emits heat, which is then used for heating. Hot water generation takes place either via the solar power system or the gas boiler (not shown here for reasons of simplicity).

During this phase, the lower heat exchanger serves as the evaporator and the zeolite as the sorption medium that takes up the refrigerant water. The adsorption phase runs until the zeolite is saturated.

Where no solar power is available, heat can also be generated using the gas burner, which, again, is not shown here in order to keep things simple.

The adsorption phase is followed by the desorption phase as described below.

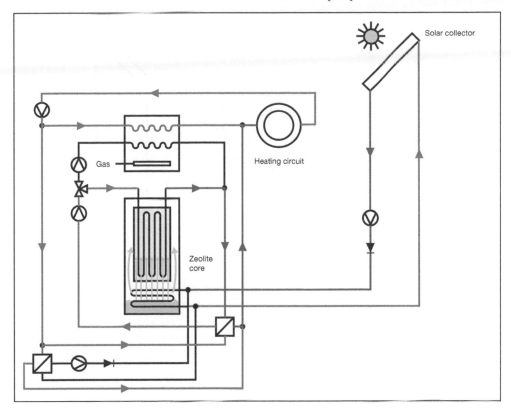

Figure 3.42 Adsorption phase of a zeolite heat pump

The desorption phase:
The desorption phase begins after all the water has been taken up by the zeolite. The gas boiler ignites in order to evaporate the water in the zeolite core. By heating the zeolite core to around 120 °C, the water is desorbed, condensing on the pipe heat exchanger in the lower section of the zeolite module, where it then collects.

During the desorption phase, the lower heat exchanger serves as a condenser, and the heat exchanger on the zeolite as a desorber.

The condensation heat is fed directly into the heating circuit. For detailed descriptions, please refer to the manufacturer's product descriptions.

Figure 3.44 shows a zeolite heat pump with a gas boiler in the upper section and the integrated zeolite module in the lower section.

A refrigeration engineer once sat next to me during a lecture in which a zeolite heat pump was presented, and he suddenly said, 'That's not a heat pump!' Is it not a heat pump? A heat pump is defined as a thermal machine that 'pumps' heat from a lower to a higher temperature. Strictly speaking, this is not the case here. Yet the zeolite heat pump uses environmental energy – namely, solar power – that can be utilised with very low flow temperatures. Gas and the gas boiler serve as an auxiliary source of energy to subsequently drive the water back out of the zeolite (desorption phase). This technique is similar to adsorption heat pumps. So we can certainly argue about the term 'heat pump'; there is no precise definition or limitation to the term.

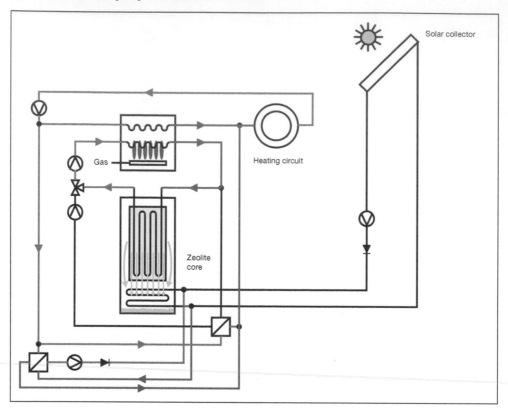

Solar collector

Gas

Heating circuit

Zeolite
core

Figure 3.43 Desorption phase of a zeolite heat pump

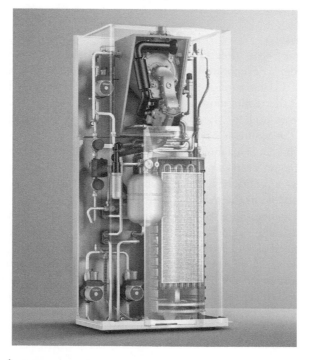

Figure 3.44 Zeolite heat pump

The energy savings associated with a zeolite gas heat pump come from the higher level of efficiency compared to standard gas boilers – the consequence of using solar power. As the technology is so new, there are currently no precise economic efficiency calculations (i.e. effort and additional cost compared to energy savings). However, the author assumes that a heat pump in the conventional sense offers better amortisation. Vaillant claims an increase in standard utilisation ratio of around 20 per cent over conventional gas boilers.

It should be noted that this is a new technology and one that will certainly be developed further. When the railways were first invented, it was believed that their speed would be dangerous for humans – and today we fly faster than the speed of sound. Therefore, we should always be open to new developments and technical improvements. This is the only way of encouraging new developments!

Zeolite heat pumps can be used in locations where, for a variety of reasons, no standard heat pump can be used, whether water-water, brine-water or air-water heat pump.

3.15 Air-air heat pumps

Air-air heat pumps are heat pumps that use the heat drawn from air to heat air. Air-air heat pumps are commonly used in air conditioning units, especially in passive houses.

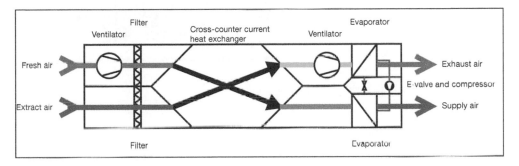

Figure 3.45 Schematic diagram of an air-air heat pump in a ventilation unit

Now we need to define the airflows:

- Fresh air: the fresh air drawn in from outside (can also be described as external air).
- Extract air: the air extracted from living areas.
- Exhaust air: the air blown outside.
- Supply air: the fresh and warm air fed into the living areas.

Here, the air-air heat pump is a component of a waste heat recovery system. Heat is extracted from the exhaust air, and the recovered heat used to warm the supply air fed into the living areas. Ventilation devices such as these are a very convenient means of heating passive houses.

Where necessary, the heat extracted from the exhaust air can also be used to heat water for domestic use. A solar power system is recommended for passive houses as a supplementary means of heating domestic water.

1   Extract air

2   Supply air

3   Control

4   Plate heat exchanger

5   Cross-counter current heat
    exchanger

6   Compressor air-air heat pump

7   Heat exchanger for extracting
    heat from the exhaust air

8   Fresh air via summer bypass

9   Fresh air

10 Exhaust air

Figure 3.46 Ventilation unit with an air-air heat pump

3.16 Other heat pump applications

Heat pumps are also used for leisure applications, as air conditioning units for heating and cooling. This offers campers the advantage of no longer needing gas for heating, and, as power costs at camp sites are often charged on a per diem basis, this allows campers to make significant savings. All that is required is the necessary power input, especially when the fridge, TV and lighting are all being used simultaneously.

Figure 3.47 Camper van with heat pump in the roof

Figure 3.48 Indoor heat pump

The heat pumps are reversible, providing warmth on cold days, and cooling on hot days. The heat source is ambient air.

The following is an example of a heat pump for commercial use, a so-called rooftop heat pump. This air conditioner has an integrated heat pump for heating and cooling buildings. An extra gas boiler can be added as required.

Figure 3.49 Rooftop heat pump

These rooftop devices can be used for the commercial heating and cooling of small and larger office buildings and business premises, commercial spaces and halls, etc., with heating and cooling outputs of up to 170 kW. The following illustrations show a few instances of their use.

Figure 3.53 shows one possible application. The circulated air and supply air are directed downwards. The fresh air is drawn in through the lateral grating, and the exhaust air expelled through the grating on the opposite side.

Figure 3.50 Gastronomy

Figure 3.51 Retail outlets

Figure 3.52 Commercial premises

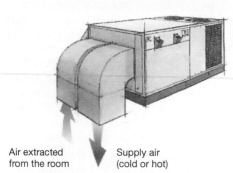

Air extracted   Supply air
from the room   (cold or hot)

Figure 3.53 Example of use

Heat pumps have many other applications. In general, where two different temperatures are needed, then it is worth considering whether a heat pump can be used effectively. Here are a few examples:

- Cooling in agricultural operations: all dairies need to cool milk. Where milk volumes are significant, it can be sensible to use the heat extracted from the milk for other purposes (e.g. hot water generation).
- Hotels, gastronomy, meat processing operations: these all involve cooling large quantities of foodstuffs. This heat could at least be used for hot water generation.
- Cooling in production plants: this often generates large quantities of heat as the result of process-related cooling. Here, too, it is worthwhile considering if this energy can be used, rather than simply emitting it, unused, into the atmosphere via water coolers or cooling towers. Where the quantity of energy is large enough, it can be used for heating, etc. Any energy that cannot be utilised can always be transferred to the atmosphere afterwards.

Planners, operational managers and technicians should all consider how and where energy can be sensibly used. This helps protect the environment and reduces costs.

4 Heat pump output

A heat pump's degree of efficiency is described by its coefficient of performance, or COP.

In heat pumps, the COP describes the heat output to electric power consumption ratio:

$$COP = P_H / P_E$$

where COP = coefficient of performance

P_H = heat pump output [kW]

P_E = electrical power consumption of the heat pump [kW]

According to German standard DIN EN 14511-2, the heat pump manufacturer is obliged to provide these specifications. Occasionally, manufacturers will give slightly different details, designed to make the COP appear better.

The coefficient of performance is also determined by the temperature of the heat source (e.g. brine) and the temperature of the emitted heat, usually water (heating water).

However, the coefficient of performance only considers the heat pump unit itself. It shows how a heat pump is significantly more efficient than other heating devices (boilers, etc.)

It is also interesting to consider the annual performance rating for the entire heat pump heating system on the basis of the seasonal performance factor β. This helps analyse the cost effectiveness of the entire heat pump system, including all the pumps, etc.

4.1 Coefficient of performance

The coefficient of performance (COP) ε is physically defined as:

$$\varepsilon = P_H / P_I$$

and $P_H = P_E + P_I$

$\rightarrow$ $\varepsilon = (P_E + P_I) / P_I$

where P_H = heat output [kW]

P_I = input power [kW]

P_E = input from the environment [kW]

The COP can also be defined according to the Carnot cycle ε_C through the temperature difference between heat source and heat utilisation, as follows:

$$\varepsilon_C = T_H / (T_H - T_A)$$

where T_H = temperature of heat utilisation = heating temperature [K]

T_A = ambient temperature of the heat source [K]

This equation shows that the COP falls significantly as the heating temperature increases. The following diagram demonstrates this for a brine-water heat pump with a brine temperature of 0 °C = 273 K.

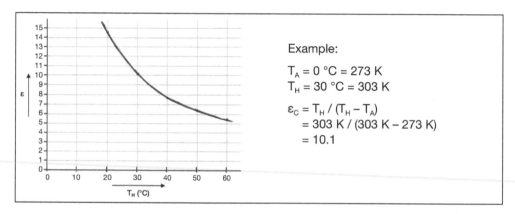

Figure 4.1 COP ε_C in the Carnot cycle

This diagram of an ideal cycle shows very clearly that as the difference in temperature between the heat output and heat source increases (i.e. flow temperatures and ground heat), the performance factor declines significantly.

The ideal cycle does not take into account losses and is therefore impossible in practice. As a result of thermal, mechanical and electrical losses, the COP for the real heat pump cycle is correspondingly lower than the COP for an ideal Carnot cycle.

In an initial approximation:

$$\varepsilon = 0.5 \times \varepsilon_C$$

In practice, this equation comes astonishingly close to reality.

There is a further method of calculating the COP, namely using cycle enthalpy. For the ideal Carnot cycle, the COP can also be determined using the h-p diagram for each refrigerant (here, R407c).

Here:

$$\varepsilon_C = (h_2 - h_3) / (h_2 - h_1)$$

where h_1 = enthalpy in front of the compressor
h_2 = enthalpy behind the compressor
h_3 = enthalpy at the injection valve

For the real cycle:

$$\varepsilon = (h_{2'} - h_{3'}) / (h_{2'} - h_{1'})$$

where $h_{1'}$ = enthalpy in front of the compressor
$h_{2'}$ = enthalpy behind the compressor
$h_{3'}$ = enthalpy at the injection valve

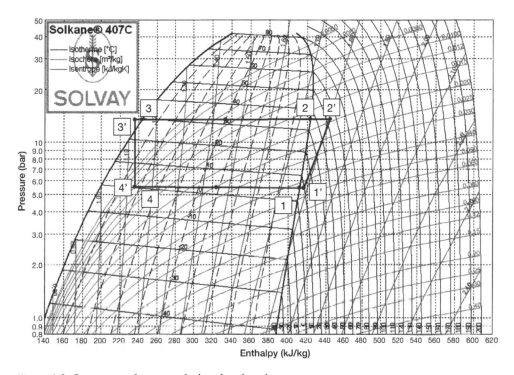

Figure 4.2 Comparison between ideal and real cycle

1 corresponds to h_1; 1′ corresponds to $h_{1'}$; 2 corresponds to h_2; 2′ corresponds to $h_{2'}$; 3 corresponds to $h_3 = h_4$; 3′ corresponds to $h_{3'} = h_{4'}$

The ideal cycle is described with the following phases:

1 – 2   Compression of the refrigerant by the compressor
2 – 3   Condensation of the refrigerant: transfer of condensation heat to the condenser
3 – 4   Decompression of the refrigerant without the release of energy
4 – 1   Evaporation, take-up of the vaporisation heat from the environment in the evaporator

In the real cycle, we need to consider the suction gas overheating and the liquid sub-cooling. For the real cycle with suction gas overheating and liquid sub-cooling, the following applies:

1' – 2'   Compression to compression temperature (superheated working vapour)
2' – 2   Cooling to condensation temperature: transfer of the superheat, $h_{2'} - h_2$
2 – 3   Condensation of the refrigerant: transfer of the condensation heat to the condenser, $h_{2'} - h_3$
3 – 3'   Liquid sub-cooling
3' – 4'   Decompression in the wet-steam region, no transfer of energy
4' – 1   Evaporation, take-up of the vaporisation heat from the environment in the evaporator, $h_1 - h_4$
1 – 1'   Suction gas overheating

According to the diagram described above, the coefficient of performance is calculated as follows:

$h_{1'}$ = 416 kJ/kg
$h_{2'}$ = 448 kJ/kg
$h_{3'}$ = 235 kJ/kg

→ $\varepsilon = (h_{2'} - h_{3'}) / (h_{2'} - h_{1'})$

= (448 kJ/kg – 235 kJ/kg) / (448 kJ/kg – 416 kJ/kg)

= 213 / 32

= 6.6

This is a very good coefficient of performance. The diagram shows that in this case, the flow temperature is around 30 °C (the condensation temperature is somewhat higher), so that the COP is correspondingly greater.

4.1.1 Coefficient of performance for a water-water heat pump

The following standard reference conditions apply for the COP during heating operations: the temperature of the well water (T_A) is 10 °C. The temperature of the water fed into the discharge well is 7 °C. The flow temperature for heating as it comes out of the heat pump is 35 °C, and the return temperature at the entrance to the heat pump is taken as 30 °C. The COP is shown as follows:

COP W10W35

'W10' represents the data on the heat source side. 'W' means 'water', here well water, and '10' the well water temperature of 10 °C. 'W35' stands for the heat output. Here again, 'W' means 'water', the heating water, and '35' the flow temperature of the heating water, 35 °C.

Again, German standard DIN EN 14511-2 states that the cooling is only 3 K (i.e. in the example above, the well water is cooled from 10 °C to 7 °C). However, in practice, this is not realistic – cooling depends on the flow rate of the underwater pump. Where this is higher, then the cooling is correspondingly less, and vice versa.

TIP

The throttle should never be used to reduce the flow rate in order to achieve the nominal flow rate.

The COP of a water-water heat pump is usually: COP (W10W35) > 5.

For the COP, the following operating reference conditions apply for hot water generation: the well water is again 10 °C. The flow temperature of the hot water leaving the heat pump is 55 °C. The temperature of the water to be fed into the discharge well and the return temperature is not determined here. The COP is then shown as:

COP W10W55

Again, this means 'W10' = (well) water temperature 10 °C, and 'W55' = flow temperature for hot water generation, 55 °C.

IMPORTANT

Often lower temperatures are given for the COP for hot water generation so that a better COP can be claimed (e.g. COP W10/W45).

4.1.2 Coefficient of performance for a brine-water heat pump

The following standard reference conditions apply for the COP of a brine-water heat pump used for heating: the temperature of the brine as it enters the evaporator is 0 °C. The temperature of the brine as it leaves the evaporator is –3 °C. The flow temperature for heating as it leaves the heat pump is 35 °C, and the return temperature as it enters the heat pump is set at 30 °C. The COP is then shown as follows:

COP B0W35

'B0' represents the data at the heat source side: 'B' is brine, and '0' is the brine temperature of 0 °C. 'W35' stands for the heat output: 'W' stands for (heating) water, and '35' = flow temperature of the heating water.

Again, DIN EN 14511-2 states that the cooling is only 3 K (i.e. for the data above, the brine is cooled from 0 °C to –3 °C). Here, too, cooling depends upon the flow rate of the brine circulation pump. Where this is higher, then the cooling is correspondingly less, and vice versa.

Thus 'B0' = brine temperature 0 °C and 'W35' = heating flow temperature 35 °C.

The COP of a brine-water heat pump is generally: COP (B0W35) > 4.

The following operating reference conditions apply for the COP for hot water generation: the brine temperature is again 0 °C. The flow temperature for hot water leaving the heat pump is 55 °C. The return flow temperatures are not determined. The COP is then shown as follows:

COP B0W55

This means 'B0' = brine temperature 0 °C and 'W55' = flow temperature for hot water generation 55 °C.

4.1.3 Comparing the COP of a water-water heat pump and a brine-water heat pump

Figures 4.3 and 4.4 compare the COP for a brine-water heat pump with that for a water-water heat pump.

These two diagrams show that the water-water heat pump has a significantly higher level of efficiency than a brine-water heat pump, especially for low flow temperatures. Therefore, it is recommended that using a water-water heat pump is at least considered. However, it is also important to examine the water quality, especially for its iron/manganese content, and its level of corrosiveness.

Furthermore, these theoretical depictions of COP are actually extremely realistic, demonstrating that theory and practice accord well.

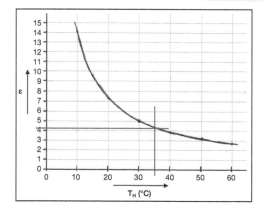

Figure 4.3
Coefficient of performance for a brine-water heat pump

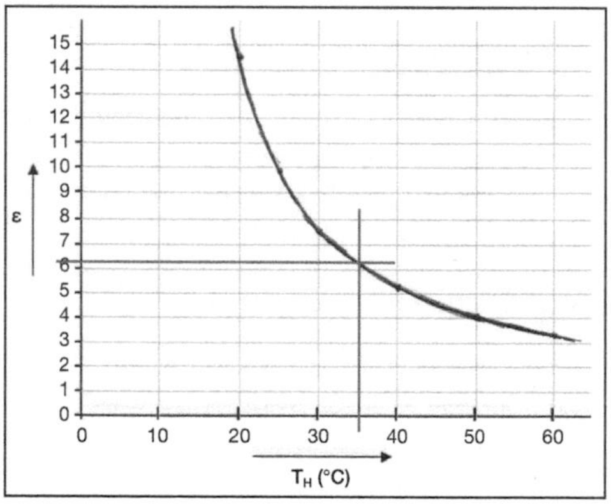

Figure 4.4
Coefficient of performance for a water-water heat pump

4.1.4 Coefficient of performance for an air-water heat pump

The following standard reference conditions apply for the COP of an air-water heat pump for heating: the air temperature (dry-bulb temperature) is 7 °C. The flow temperature of the heating water as it leaves the heat pump is 35 °C, and the return temperature on re-entering the heat pump is taken as 30 °C. The COP is shown as follows:

COP A7W35

'A' represents the data on the heat source side, where 'A' is air, and '7' is 7 °C. 'W35' represents the heat output, where 'W' is again (heating) water and '35' the heating water flow temperature 35 °C.

The COP of an air-water heat pump is usually: COP (A7/W35) < 4.

The following operating reference conditions apply to the COP for hot water generation: the air temperature is again 7 °C. The flow temperature of the water leaving the heat pump to generate hot water is 55 °C. The return temperatures are not given. The COP is then shown as follows:

COP A7W55

Here, 'A7' = air temperature 7 °C and 'W55' = flow temperature for hot water generation 55 °C.

Several COPs are given for air-water heat pumps, each for different external temperatures. In accordance with DIN EN 14511-2, a distinction is made between standard reference conditions and operating reference conditions. The standard reference conditions assume an external temperature of 7 °C. The operating reference conditions are –7 °C or 2 °C. According to German guideline VDI 4650, in order to calculate the seasonal performance factor (SPF), we also need the COP for an external temperature of 10 °C.

Thus the following COP values apply for heating:

COP A-7/W35
COP A2/W35
COP A7/W35
COP A10/W35

And for hot water generation:

COP A-7/W55
COP A2/W55
COP A7/W55
COP A10/W55

The following apply for air-air heat pumps:

COP A-7/A45
COP A2/A45
COP A7/A45
COP A10/A45

Air-air heat pumps are often used in ventilation equipment for heat recovery and additional air heating.

4.2 The seasonal performance of various heat pumps

This section will examine the performance of a water-water heat pump system, a brine-water heat pump system and an air-water heat pump system, over the course of a year. Here, the heat source – whether water, ground, brine or air – plays a key role and, in turn, has an impact on the performance factor and operating costs.

The following considerations are based on a detached family home with underfloor heating, including hot water generation and an annual heating demand of around 12,000 kWh/a. The heat is required as follows:

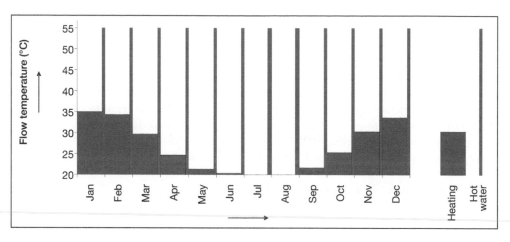

Figure 4.5 Heat demand profile

The red columns show the temperature level needed for heating, and the blue columns the temperature level for hot water generation. The column areas are a rough indication of the required heat quantities.

4.2.1 The seasonal performance of a water-water heat pump system

A water-water heat pump has two decisive advantages:

1 At 10 °C, the temperature level of the heat source is quite high.
2 The source temperature is fairly constant.

As the source temperature is almost constant, so is the efficiency of the heat pump. The COP ε here is dependent above all on the flow temperature. This is described in the following equation:

$$\varepsilon = 0.5 \times T_A / (T_H - T_A)$$

$$= \varepsilon$$

$$= f(T_H)$$

$$= 0.5 \times 283 \ K / (T_H - 283 \ K)$$

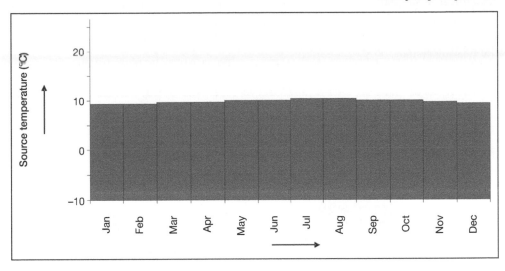

Figure 4.6 Seasonal variations in groundwater temperature

where   ε = efficiency factor
T_A = ambient heat source temperature
T_H = flow temperature for heating
$f(T_H)$ = mathematical: function of T_H

This equation demonstrates how the efficiency of a water-water heat pump depends on the flow temperature (see Section 4.1, 'Coefficient of performance').

The slightly fluctuating COPs lead to fluctuating performances for each operating state. For the performance factors, the COP for heating and hot water generation are considered each according to their share: in winter, heating dominates, and the share of hot water generation is relatively small. Therefore, the efficiency factor is also relatively high in winter. In summer, the heat pump must mainly operate with relatively high flow temperatures for

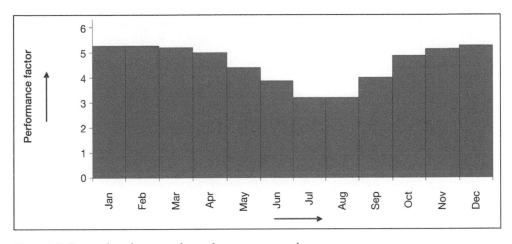

Figure 4.7 Seasonal performance factor for a water-water heat pump

hot water generation, and therefore the efficiency factor is significantly lower in summer. A solar-powered hot water generation system could improve the overall performance of the heating system.

The power consumption varies accordingly:

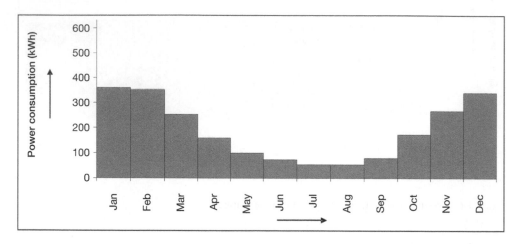

Figure 4.8 Seasonal variations in power consumption for a water-water heat pump

Unsurprisingly, the water-water heat pump requires more power in winter to cover the increased demand for heating. Due to the high, and almost constant, source temperature, the performance factor in winter is fairly high. That helps save energy and reduce costs. In summer, only a little power is required to operate the heat pump for hot water generation, and this can be further reduced with the aid of a solar power unit.

4.2.2 The seasonal performance factor of a brine-water heat pump system

The source temperature of a brine-water heat pump is not constant:

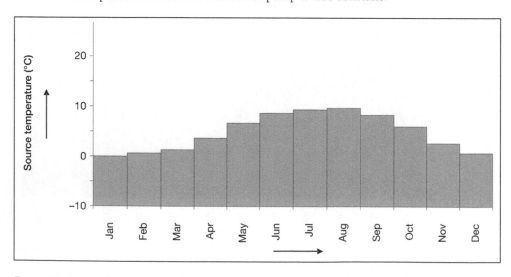

Figure 4.9 Seasonal variations in brine temperature as it enters the brine-water heat pump

According to DIN EN 14511-2, the COP for heating is defined as B0 W35. However, experience shows that the brine source temperature fluctuates significantly over the course of the year. This is because significant quantities of heat are extracted from the ground in the winter months. Consequently, the ground becomes much cooler, and the effect is compounded in ground heat exchangers installed close to the surface. During the warmer seasons, less heat is extracted from the ground, allowing the ground to regenerate as a heat source. Ground heat exchangers installed close to the surface regenerate faster in summer than do deeper borehole heat exchangers. However, in winter they cool down more, due to their proximity to the frost line.

As the source temperature varies with the seasons, the level of efficiency fluctuates more. Here, the efficiency factor depends upon the temperature of the heat source (brine) and the flow temperature. This is described in the following equation:

$$\varepsilon = f(T_H, T_A)$$
$$= 0.5 \times T_A / (T_H - T_A)$$

In mathematical terms, the coefficient of performance ε is a function of the source temperature T_A and the flow temperature T_H.

The COP, and therefore also the performance factor, is dependent on the fluctuating heat source temperature and the required flow temperature. Consequently, the efficiency factor, as expressed in the performance of a brine-water heat pump system, is subject to significant fluctuations:

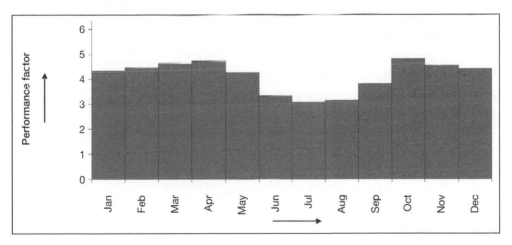

Figure 4.10 Seasonal performance for a brine-water heat pump

During the winter months, performance falls as the ground cools. Consequently, in spring, the performance factor improves as the ground begins to warm up. After regeneration in summer, the performance is somewhat higher. During the summer period itself, the performance factor is lower because here the heat pump is primarily working to generate hot water with hot flow temperatures.

This dynamic naturally has an impact on power consumption:

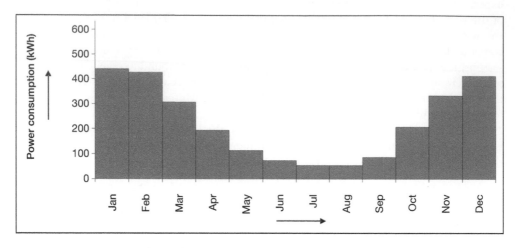

Figure 4.11 Seasonal variations in power consumption for a brine-water heat pump

Due to the lower source temperature of the brine, the power consumption of a brine-water heat pump in winter is around 20 per cent higher than for a water-water heat pump. When the ground has regenerated over the summer, then the efficiency factor of both heat pumps is almost identical – except that they are not used for heating in summer. Over the year as a whole, a brine-water heat pump consumes around 16 per cent to 18 per cent more electricity than a water-water heat pump.

4.2.3 The seasonal performance factor of an air-water heat pump system

In an air-water heat pump, the source temperature (i.e. the ambient air) is subject to even greater fluctuations:

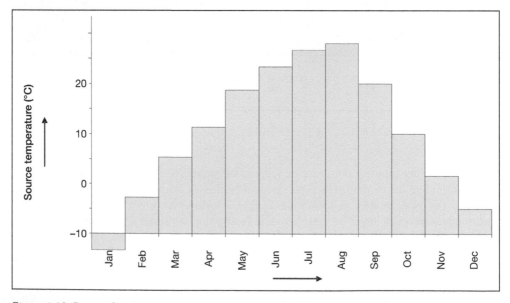

Figure 4.12 Seasonal variations in source temperature (air) for an air-water heat pump

As in the brine-water heat pump, the COP, and therefore the efficiency factor, are subject to significant fluctuations. Here again:

$$\varepsilon = f(T_H, T_A)$$
$$= 0.5 \times T_A / (T_H - T_A)$$

TIP

Here, it is also clear that where an air-water heat pump has a monovalent configuration (not monoenergetic), then this must be set for the coldest possible external temperature. Otherwise, only monoenergetic operation with the support of a heating element is possible. As external temperatures rise, the performance factor naturally increases. Therefore, a buffer tank is highly recommended for an air-water heat pump in order to avoid overly frequent cycles.

As the source temperature (i.e. the air), is subject to significant fluctuations, the COP and level of efficiency fluctuate correspondingly. This is expressed in a significantly more fluctuating performance factor:

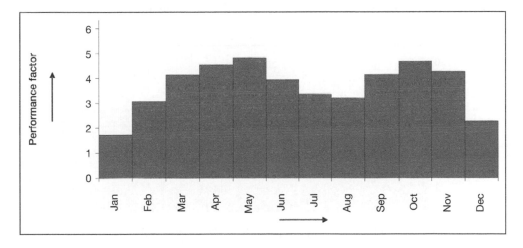

Figure 4.13 Seasonal performance for an air-water heat pump

Due to the low external temperatures in winter, the performance factor and efficiency level of an air-water heat pump are correspondingly low. In contrast, in summer the source temperatures may exceed the maximum permitted air temperatures for the heat pump, so that, for technical reasons, the heat pump can no longer be used for hot water generation. Lower performance factors, especially in winter, naturally have a negative impact on power consumption and thus electricity costs:

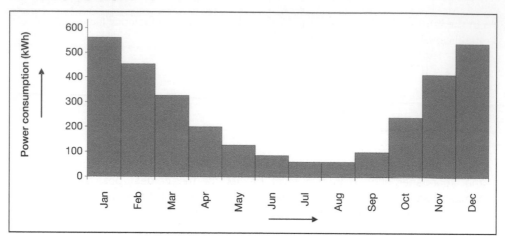

Figure 4.14 Seasonal variations in power consumption for an air-water heat pump

Here, it is clear that when the greatest amount of heat is available for heating, the performance factor is at its worst, and thus power consumption is correspondingly higher.

Most air-water heat pumps have a bivalent configuration (i.e. monoenergetic operation): on particularly cold days, they switch on the electrical heating element to support the heat pump. This reduces the performance factor during the winter months:

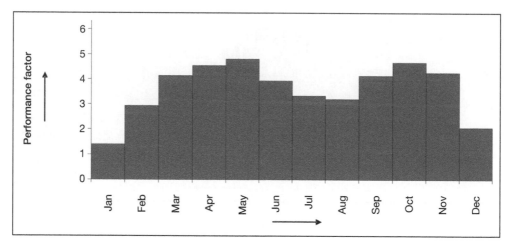

Figure 4.15 Seasonal variations in performance factor for an air-water heat pump with electrical heating element

This naturally leads to greater power consumption.

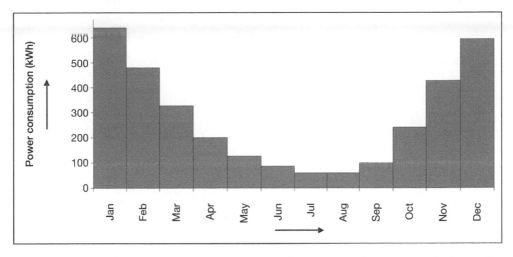

Figure 4.16 Seasonal variations in power consumption for an air-water heat pump with electrical heating element

A comparison with Figure 4.14 shows that power consumption during the cold winter months is notably higher than during the remaining months.

4.2.4 A comparison of the seasonal performance of the various heat pump systems

Our previous observations show that optimal performance is achieved with constant and high source temperatures and low flow temperatures. To illustrate this, the following figures show the temperature variations and their impact. First, a comparison of the source temperatures:

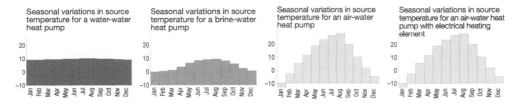

Figure 4.17 Seasonal variations in source temperature

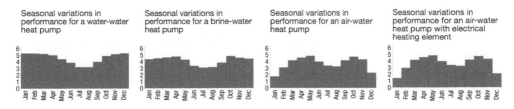

Figure 4.18 Seasonal variations in performance

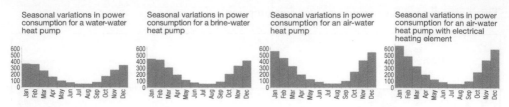

Figure 4.19 Seasonal variations in power consumption

A comparison of the various heat pump systems shows that performance factors are higher with the higher source temperatures offered by water, brine (ground) or air. The best heat source is groundwater, assuming the framework conditions are suitable. Where this is not possible, then borehole or ground heat exchangers are certainly a good alternative. The performance of brine-water heat pumps is, however, a little lower than for water-water heat pump systems. In terms of acquisition costs, air-water heat pumps are a less expensive alternative, but a compromise is made in the form of lower performance. This naturally leads to higher operating costs, especially where the air-water heat pump operation is bivalent (i.e. not monovalent, but only monoenergetic).

4.3 Seasonal performance factor

The seasonal performance factor (SPF) β describes the relationship between the overall annual heating demand (Q) of the building in question with the overall electrical power input (E), including all electrical equipment associated with the heating system (heat pump with controller, primary and secondary pumps, etc.).

- The building's annual heat demand is displayed in the thermal insulation certificate as required by the German Energy Saving Ordinance (EnEV). It comprises the annual heating requirement and the annual heat demand for heating and hot water generation.

According to the European heat pump association EHPA, the following benchmark values reflect the current state of technological development:

- Water-water heat pump systems: 4.5
- Direct evaporator heat pump systems: 4.2
- Brine-water heat pump systems: 4.0
- Air-water heat pump systems: 3.5

Significantly higher values can be achieved by optimising the systems (larger borehole/ground heat exchangers, lower flow temperatures, lower hot water temperature, etc.).

The seasonal performance factor is also influenced by the type of building, and therefore cannot be determined by the heat pump manufacturers. Instead, it must be calculated by the planner, energy consultant or engineer installing the heat pump system.

TIP

The calculated seasonal performance factor is a theoretical value that can deviate significantly from the actual seasonal performance factor, according to consumption patterns, weather conditions, etc.

4.3.1 The seasonal performance factor of a water-water heat pump system

The seasonal performance factor is calculated as:

$$\beta = Q / E$$

where β = seasonal performance factor
Q = annual heat demand of the building [kWh/a]
E = annual energy consumption for all electrical components in the heat pump system [kWh/a]

The annual heat demand is calculated as the sum of the annual demand for heating and the annual heat required for hot water generation:

$$Q = Q_H + Q_{HW}$$

where Q_H = annual heating demand [kWh/a]
Q_{HW} = annual heat required for hot water generation [kWh/a]

Where the annual heat demand for hot water generation is not stated in the thermal insulation certificate, then this can be taken as 20 per cent to 25 per cent of the annual heating demand. It should be noted that this depends on the relationship between the size of the living space being heated and the number of residents.

The annual energy consumption for all the electrical components in the heat pump system is calculated as:

$$E = E_{HP} + E_{UP}$$

where E_{HP} = annual energy consumption for the heat pump [kWh/a]
E_{UP} = annual energy consumption for the primary pump (underwater pump) [kWh/a]

The annual energy consumption for the heat pump is calculated as:

$$E_{HP} = Q_H / COP_H + Q_{HW} / COP_{HW}$$

where COP_H = COP for heating operation W10W35
COP_{HW} = COP for hot water generation W10W55

Attention COP$_H$!

The COP$_H$ is unequal to COP W10W35, because the flow temperature of 35 °C relates to the minimum dimensioning temperature (e.g. –10 °C). As controllers regulate according to the weather conditions, the actual COP$_H$ is dependent on the flow temperature, which in turn depends on the external temperature. Thus:

$$COP_H = f(T_H)$$
$$= \varepsilon$$
$$= C \times T_A / (T_H - T_A)$$

where $COP_H = f(T_H) = COP$ as function of flow temperature

C = constant, roughly 0.5

The flow temperature falls as the external temperature rises, with the COP rising correspondingly. The average COP_H over the entire year is certainly significantly larger. Assuming, for example, an average flow temperature of 30 °C, then the COP_H improves by around 25 per cent, and with an average flow temperature of only 25 °C, by as much as 65 per cent. The runtimes must also be taken into account; naturally, they are reduced as external temperatures increase.

Attention COP_{HW}!

It is not absolutely necessary, and therefore makes no sense, to constantly heat the hot water tank to 55 °C. Often, 45 °C is sufficient. This improves the COP by around 24 per cent. Here, the COP_{HW} is calculated as follows (cf. calculation of ε):

$$COP_{HW} = COP_{W10W45}$$
$$= 1.24 \times COP_{W10W55}$$

The annual electrical energy consumption for the underwater pump is calculated as follows:

$$E_{UP} = P_{UP} \times t_{UP}$$

where P_{UP} = electrical power of the underwater pump [W/kW]
    t_{UP} = annual operating time of the underwater pump [h]

and $P_{UP} = (P_H + P_L) \times Q_{HP} / \cos \varphi$

where P_H = pressure loss caused by height difference between groundwater level and
        heat pump [Pa] = [kg/ms²]
    P_L = pressure loss in the PE pipes running to and from the heat pump [Pa] = [kg/ms²]
    Q_{HP} = volume flow of the well water through the heat pump [m³/h]
    $\cos \varphi$ = phase angle of the underwater pump

Conversion aid: pressure loss for 10 m = 1 bar = 10^5 Pa = 100,000 Pa

In a first approximation, we can assume:

$$t_{UP} = t$$

where   t = system operating time [h]

$$t_{UP} = t = Q_H / P_{HPH} + Q_{HW} / P_{HPHW}$$

where P_{HPH} = heat output of the heat pump for heating at W10W35 [kW]

P_{HPHW} = heat output of the heat pump for hot water generation at W10W55
        [kW]

These theoretical observations are designed to help give a feeling for the thermodynamic correlations; however, they cannot be used as such in practice.

4.3.2 The seasonal performance factor of a brine-water heat pump system

The seasonal performance factor is calculated as:

$$\beta = Q / E$$

where β = seasonal performance factor
   Q = annual heat demand of the building [kWh/a]
   E = annual energy consumption for all electrical components in the heat pump system [kWh/a]

The annual heat demand is calculated as the sum of the annual demand for heating and the annual heat required for hot water generation:

$$Q = Q_H + Q_{HW}$$

where Q_H = annual heating demand [kWh/a]
   Q_{HW} = annual heat required for hot water generation [kWh/a]

The annual energy consumption for all the electrical components in the heat pump system is calculated as:

$$E = E_{HP} + E_{PP}$$

where E_{HP} = annual energy consumption for the heat pump [kWh/a]
   E_{PP} = annual energy consumption for the primary pump (brine pump) [kWh/a]

The annual energy consumption for the heat pump is calculated as:

$$E_{HP} = Q_H / COP_H + Q_{HW} / COP_{HW}$$

where COP_H = COP for heating operation B0W35
   COP_{HW} = COP for hot water generation B0W55

Attention COP_H!
The COP_H is unequal to COP B0W35, because the flow temperature of 35 °C relates to the minimum dimensioning temperature (e.g. −10 °C). As controllers regulate according to the weather conditions, the actual COP_H depends on the flow temperature, which in turn depends on the external temperature. Furthermore, annual fluctuations in brine temperature must also be considered in a brine-water heat pump. Thus:

$$COP_H = f(T_H, T_A) = \varepsilon = C \times T_A / (T_H - T_A)$$

where COP_H = $f(T_H, T_A)$ = COP as function of flow temperature and brine temperature
   C = constant, roughly 0.5.

The flow temperature falls as the external temperature rises, and the COP rises correspondingly. Furthermore, the brine temperature fluctuates. At the beginning of a heating period, this is correspondingly higher, falling as the ground cools. At the end of the heating period, thermal regeneration of the ground begins, and the increasing warmth of the ground leads to a rise in brine temperature. Here again, the average COP_H over the entire year is certainly significantly larger.

For hot water generation, the same observations apply as for a water-water heat pump – the only difference is that the brine temperature varies.

Again, these observations cannot be directly translated into practice.

4.3.3 The seasonal performance factor of an air-water heat pump system

In Section 4.3.1, 'The seasonal performance factor of a water-water heat pump system', and Section 4.3.2, 'The seasonal performance factor of a brine-water heat pump system', we see that the seasonal performance factor calculations are almost identical. In principle, they also apply to the air-water heat pump. In an air-water heat pump, the heat source temperature T_A is subject to greater variations, making the thermodynamic correlations even more complex.

As the evaporator fan is already accounted for in the heat pump performance, only the electrical consumption of the secondary pump needs to be considered.

4.3.4 Calculating the seasonal performance factor in accordance with German guideline VDI 4650

In the newly revised VDI 4650 Part 1 (2009), the seasonal performance factor (SPF) calculation is regulated and applicable for electrical heat pumps for space heating, including hot water generation for all standard heat pump systems. Therefore, the following is only a small excerpt from the new VDI 4650 Part 1. In the new VDI 4650 Part 1, a differentiation is made between planning and subsequent inspection. Examples will clarify this. The VDI guideline still serves to calculate the SPF, to calculate the expected savings, and to inspect the actual savings made with existing systems. It should be noted that standard values are used during planning, and these can vary significantly in the final inspection as nature fails to be guided by standard values.

TIP

The SPF β is specific to the type of building and system, and therefore cannot be calculated by the heat pump manufacturer, only by the installing engineer. Even so, manufacturers often calculate the SPF as an aid to planning.

External influences such as weather conditions can lead to significant deviations between the calculated and actual seasonal performance factor.

The SPF is calculated as:

$$\beta = 1 \,/\, e_{HP}$$

where β = seasonal performance factor
e_{HP} = annual power input factor

and $e_{HP} = F_P \,/\, (\varepsilon_N \times F_{\Delta\theta} \times F_\theta)$

where F_P = correction factor to account for heat pump source
ε_N = coefficient of performance of the heat pump (see the heat pump's technical details)
$F_{\Delta\theta}$ = correction factor for deviations in brine temperature
F_θ = correction factor for differing operating conditions

With regards to the correction factors, the tables listed in the VDI 4650 Part 1 apply. For the sake of simplicity, the following values are given:

F_P = 1.075 for planning a brine-water heat pump
$F_{\Delta\theta}$ = 1
F_θ see Table 4.1 (e.g. for a brine-water heat pump system)

Table 4.1 Correction factors F_θ for brine-water heat pumps under different operating conditions

$T_{Brine,min}$ [°C]	Max flow temperature $T_{flowtemp.max}$ [°C]					
	30	35	40	45	50	55
2	1.161	1.113	1.065	1.016	0.967	0.917
1	1.148	1.100	1.052	1.003	0.954	0.904
0	1.135	1.087	1.039	0.990	0.940	0.890
−1	1.122	1.074	1.026	0.977	0.927	0.877
−2	1.110	1.062	1.014	0.965	0.915	0.864
−3	1.099	1.051	1.002	0.953	0.903	0.852

The VDI 4650 Part 1 also includes tables with correction factors for water-water heat pumps, direct evaporator heat pumps and air-water heat pumps.

In the current VDI 4650 Part 1, cooling is not considered when determining the SPF. However, its impact on the SPF should not be ignored, and can lead to significant deviations, especially when a heat pump operates in reverse.

4.4 Primary energy factor fp and energy input factor e_p

The energy input factor e_p, often referred to as the system expenditure, is the inverse of the seasonal performance factor, multiplied by the primary energy factor of the electrical energy:

$$e_p = fp / \beta$$

where   fp = energy factor or primary energy factor
   β = seasonal performance factor

fp = 2.6 for the energy factor of the electrical energy. This is the average value for all supplying power plants (coal-fired and nuclear power, gas and oil-fired power plants, combined heat and power plants, wind and solar parks, etc.), and including transmission losses, although these are almost negligible.

In determining the energy input factor e_p, the entire heating system including hot water and energy generation (electricity generation and transport) is evaluated. This is regulated in German standard DIN 4701-10.

The energy input factor e_p is an ecological consideration and describes the relationship between the primary energy demand (oil, natural gas, coal, etc.) and the heat generated by the heating system. The smaller the energy input factor e_p, the more effective and environmentally friendly the heating system.

For example, where the energy input factor e_p is 1.5, then this means that one and a half times more energy is consumed by the heating system than the overall heat output. When calculating the energy input factor e_p, all the primary energy required to generate the heat energy is considered, including its extraction and transport, conversion, heat distribution, transmission losses, control losses, etc.

Usually, the heat pump performs significantly better than most other heating systems. This is shown in the following sample calculation, which compares two identical houses, one with a condensing gas boiler and the other with a heat pump. The heat demand is 12,000 kWh/a.

Table 4.2 The energy factors for various heat generators

Electrical heating	fp = 2.6
Night storage heating	fp = 2.0
Lignite-fired heating	fp = 1.2
Oil heating	fp = 1.1
Gas heating	fp = 1.1
Coal-fired heating	fp = 1.1
District heating from thermal power plants using fossil fuels	fp = 1.1 to 1.3*
District heating from combined heat and power plants using fossil fuels	fp = 0.2 to 0.9*
Wood heating	fp = 0.2
District heating from thermal power plants using renewable energy	fp = 0.0 to 0.2*
District heating from combined heat and power plants using renewable energy	fp = 0.0
Renewable energy	fp = 0.0

* Depending upon the method of generating and transmitting energy

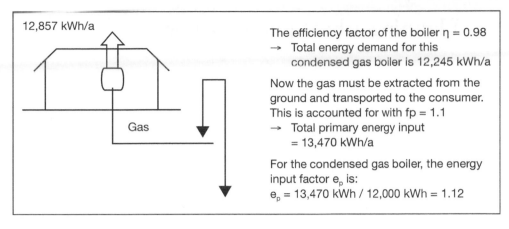

The efficiency factor of the boiler η = 0.98
→ Total energy demand for this condensed gas boiler is 12,245 kWh/a

Now the gas must be extracted from the ground and transported to the consumer. This is accounted for with fp = 1.1
→ Total primary energy input = 13,470 kWh/a

For the condensed gas boiler, the energy input factor e_p is:
e_p = 13,470 kWh / 12,000 kWh = 1.12

Figure 4.20 House with modern condensing gas boiler

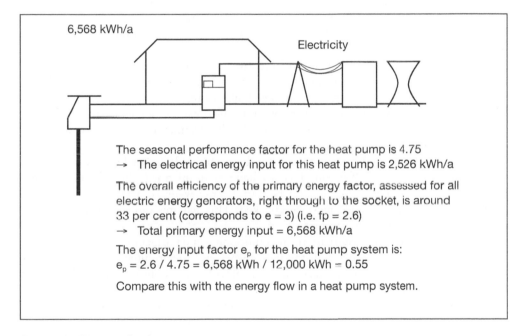

The seasonal performance factor for the heat pump is 4.75
→ The electrical energy input for this heat pump is 2,526 kWh/a

The overall efficiency of the primary energy factor, assessed for all electric energy generators, right through to the socket, is around 33 per cent (corresponds to e = 3) (i.e. fp = 2.6)
→ Total primary energy input = 6,568 kWh/a

The energy input factor e_p for the heat pump system is:
e_p = 2.6 / 4.75 = 6,568 kWh / 12,000 kWh = 0.55

Compare this with the energy flow in a heat pump system.

Figure 4.21 House with a heat pump system

Only very badly planned heat pump systems have a seasonal performance factor of 3 and thus an energy input factor of e_p = 1.

Any heating system not run using renewable energy increases the quantity of CO_2 in the atmosphere. In these terms, the only winners are wood or pellet heating systems, which operate with a closed CO_2 cycle. Or a heat pump system driven using photovoltaics. So why not green electricity?

Therefore, from an ecological and economic perspective, a heat pump run using green electricity is the best type of heating system.

5 Planning a heat pump system

Planning a heat pump system is a cooperative venture involving architects, specialist planners, civil engineering companies, heating engineers and electricians, as well as the eventual homeowners.

Prior to the start of planning, enquiries should be made with environmental agencies and local water authorities, checking which heat pump systems are permitted at the chosen location. This can vary considerably from location to location. Environmental aspects and water laws must all be considered.

First, the architect needs to plan a sufficiently large boiler room in coordination with the owners of the planned heat pump system. In principle, the boiler room should offer sufficient space for a heat pump system with two tanks. Compact heat pumps with integrated hot water tanks require less space and are often installed without a buffer tank. The space-saving advantage needs to be weighed against the disadvantages of spare parts purchases and maintenance costs.

There must also be sufficient space for the heat source. It is usually easiest to install borehole heat exchangers. If ground heat exchangers are planned, then the first consideration must be the size of the plot of land. For a water-water heat pump, the ability to maintain a minimum distance must be checked. This requires discussions between the architect and each specialist contractor.

The planner for heating, ventilation, air conditioning and refrigeration (HVACR) plans the heat pump system together with the homeowner and the architect, and takes care of the technical execution of the project. Any special wishes the homeowner may have must be taken into account.

For a heat pump using wells or borehole heat exchangers, the well engineer must be involved in deciding where the heat source should be installed. The engineer will need documentation and overview maps issued by the local geological service when planning the position of the wells, and in particular the borehole heat exchangers. There must be sufficient access for any heavy drilling equipment. Consequently, it often makes sense to install the wells or borehole heat exchangers prior to the start of house construction; water may even be drawn from the wells for use during house construction.

In the end, the heating engineers and electricians install the heat pump system in accordance with the details provided by the specialist planner or manufacturer.

Careful and precise planning is extremely important when it comes to heat pump systems. First, the output of the heat pump must be calculated. Insufficient output often leads to the electric heating element – if present – being switched on. This naturally eliminates the advantages of a heat pump's high level of efficiency. Therefore, the heat pump output must be sufficiently generous so that the electric heating element is not necessary, only serving for emergencies. In principle, a heat pump should be operated in monovalent mode. This means that the heating demand (heating + hot water) is exclusively covered by the heat pump. In bivalent operations, the heat pump has an alternative heating source (oil or natural gas). From an economic point of view, this makes no sense.

Bivalent heat pumps cover the heating demand up to the so-called bivalence point. The bivalence point corresponds to the ambient temperature at which point a second heat generator is switched on. This is often the case for an air-water heat pump (see Section 5.5.3, 'Example of planning an air-water heat pump system').

Heat pump systems with supplementary electrical heating are often described as monoenergetic heat pump systems. This means that only one energy source – electricity – is required to operate the system. However, this by no means excludes direct electrical auxiliary heating or even direct electrical heating (e.g. for generating hot water using an electric heating element). And this can be expensive!

PLEASE NOTE

Be careful! Monoenergetic is not the same as monovalent.

The next step is to decide between a water-water or a brine-water heat pump.

A heat pump system should be selected according to the customer's wishes and with regard to the local geological characteristics.

5.1 The tasks of an architect and consultant

Architects, planners and property developers are usually the first point of contact for potential homeowners. Therefore, they play a very important role when it comes to providing advice on energy-saving measures, including insulation, using renewable energy and heat pumps. This will be increasingly important in future as heat pumps are being more frequently installed, especially in new builds. Therefore, a well-informed architect must have at least basic knowledge of heat pumps to ensure that the homeowner receives the appropriate advice, planning and construction supervision.

This chapter, directed at planners and architects, only touches on the key points involved in advising, planning and installing a heat pump system. Further information is provided in the remaining chapters of this book.

5.1.1 Advice and planning from the architect and consultant

After the initial contact and discussions with the homeowner, the first step is planning. A well-informed architect should also know how large the boiler room should be, because a complete heat pump system consists of a heat pump, a hot water tank and a buffer tank. These system components require more space than a wall-hung condensing boiler with a small standing hot water tank beneath. Naturally, space-saving compact heat pump systems are also available – heat pump units with integrated hot water tanks, often available without buffer tanks. Here is a little story about these compact units: I recently received a phone call from a customer who had an older compact heat pump with an integrated hot water tank that was leaking. The heat pump itself was still working. However, replacements for the leaking tank on the older compact unit were no longer available. The customer considered a new heat pump system with individual components. But he did not need to think for long because he just did not have the space. And thus the compact unit ended up on the scrap heap, which was a shame. And the customer was forced to buy a new compact heat pump. That was expensive! And what happens in 15 years' time? This example clearly shows the importance of competent and fair advice right from the start.

Unsurprisingly, many property developers focus on immediate profits, and these can be promptly and easily achieved by offering compact units that are quick to install. But why do they not offer potential homeowners an alternative? For example, instead of a compact heat pump, they should offer a heat pump system made up from individual components, presenting potential buyers with a list of all the advantages and disadvantages. Rather than making a quick sale, the developer can sell 'added value' – and earn more by doing so. However, this does require a solid understanding of heat pumps.

Furthermore, a well-informed architect and consultant should be well versed in the various types of heat pump (e.g. water-water, brine-water or air-water heat pumps) and understand the circumstances for which each is suited. For example, it is not appropriate to recommend an air-water heat pump in the form of an outdoor unit in a new and tightly packed residential area. This could give rise to arguments with neighbours and create social tensions. This does not have to be the case.

Homeowners wanting a heat pump are intending to save money on heating. Therefore, they want the most efficient heat pump available. Water-water heat pumps are the most efficient, but it must first be determined whether the hydrological conditions and water quality make them appropriate for the location in question. Here, a good alternative would be a brine-water heat pump. Cost-effective solutions are brine-water heat pumps with ground heat exchangers, but only where sufficient space is available. This is frequently not the case for many new builds. An alternative to ground heat exchangers is borehole heat exchangers, which offer the added benefit of free cooling. Here again, specialist knowledge regarding abstraction capacity and the costs of installing borehole heat exchangers is helpful. Where the homeowner's financial resources are insufficient, then the alternative is an air-water heat pump.

An architect offers neutral, disinterested advice and planning, and therefore the information and advice provided by an architect is regarded in a different light to that offered by an installation engineer with a personal interest in securing a contract. This is not the case for an architect or planner.

5.1.2 Advice and planning from the architect and consultant, and special requirements

The architect and planner can consider any special requirements the homeowner may have during a discussion. The options are many and various, and here we touch on a few.

5.1.2.1 Heating swimming pools

Here, we need to differentiate between indoor and outdoor pools. Outdoor pools are usually only heated in summer. Here, we need to determine if more heat is needed to heat the pool or the building. The heat pump output will be determined accordingly.

It is assumed that an indoor swimming pool will be used throughout the year. There is also the input needed to heat the building, as well as hot water generation for use in the building and for heating the swimming pool. A single heat pump that covers all these demands is recommended, rather than one for the building and a second for the swimming pool. For indoor swimming pools, consideration must also be paid to dehumidifying the hall in which the pool is located, and the heat loss this involves.

5.1.2.2 Heating outbuildings

Heating outbuildings such as garages or greenhouses must be considered separately. The heat input required here should be calculated separately and then added to that needed for the main building. Here, there are a range of possibilities:

$$P_{Overall} = P_H + P_{Outbuildings}$$

where $P_{Overall}$ = overall heating requirement
P_H = heat output for the building
$P_{Outbuildings}$ = heating for outbuildings

Here, the homeowner's wishes must be considered, in accordance with statutory requirements.

5.1.3 Monitoring and supervision of construction by the architect and consultant

If a planner, architect or site manager wishes to supervise construction, then they need the relevant specialist knowledge and experience.

TIP

The heat source (i.e. well, borehole or ground heat exchangers) should be installed prior to the start of construction.

Installation at a later date can cause unnecessary expense. Wells drilled for a water-water heat pump can be used to provide water needed during construction. This removes the need for a water hydrant and the cost of water. It is also recommended that borehole or ground heat exchangers are installed prior to building. When borehole heat exchangers are installed at a later date, the well engineer may have difficulty manoeuvring heavy drilling equipment when the ground for digging and earthworks for connections (water, sewage, rainwater, electrical cables, etc.) is no longer homogenous and uniformly solid. Costs can also be saved if the borehole heat exchangers are installed prior to building. If ground heat exchangers are to be installed when the house is almost complete, then the company installing the pipes has a problem with where to place the excavated soil during installation. Where does this go when the plot is too small, as is often the case? Borehole and ground heat exchangers installed prior to construction do not disrupt building construction in the slightest. In contrast, their subsequent installation can give rise to higher costs where building conditions are complicated.

When building the foundations (i.e. the foundation slab or cellar), the necessary reserve conduits and wall penetrations can be planned in advance. If installed at a later date, these conduits and wall penetrations add to the building costs. Accordingly, the installation of any future brine circuit manifold, either inside or outside the building, should also be planned ahead. Within the building, thought must be given to condensation build-up. Outside the building, a distribution shaft is required for the installation of a brine circuit manifold. It should be accessible at all times.

The heat source (i.e. intake and discharge wells, borehole and ground heat exchangers) must be registered with the local water authorities prior to installation, and legal permission obtained with respect to the geological and hydrological requirements. The well engineer and installation engineer must first obtain the necessary data from the local water authorities. This is a service often undertaken by the well and/or installation engineer. However, who carries this out is less important than the fact that this step is completed.

The electrician must ensure that sufficient power is available to operate the heat pump. They must research the conditions offered by the local power utility. Where the heat pump is relatively large, it may make sense to use the utility's cheaper tariffs. For smaller heat pumps, this may bring no benefit. In each case, the size of the heat pump and the available tariffs must be considered. Electricians should be aware that using a heat pump tariff requires the use of a larger meter board for which there will need to be sufficient space.

Architects and planners must consider the installation process itself. The heat pump and its necessary components should be ordered prior to installing the underfloor heating; the heat pump should be in place in time to heat the floor screed and the new building. When it comes to the floor screed heating programme and heating to dry out the building, the season must also be considered. Thanks to recently amended German energy-saving regulations stipulating increased insulation levels, the required heat source output continues to decrease. Consequently, for new builds with a high water content (e.g. brick, lime sand brick, etc.) the heat pump output may be insufficient to dry the floor screed or for dry heating, especially in winter. Planning should also include a construction dryer or additional electrical heating.

5.2 Determining heat pump output

First, we need to differentiate between a new build and existing housing stock. New builds are very easy to plan, and here the relevant laws for calculating heat output apply.

For existing housing, the output of the existing boiler should not be used as a guide as these are often significantly oversized. It is recommended that an energy consultant or engineering office be instructed to provide a precise calculation. Alternatively, it may make sense to consider modernising the building (energy-saving measures) before planning the heat pump system.

5.2.1 Determining heat pump output for a new build

Calculating heat output

According to the German energy-saving ordinance (EnEV) 2002, new builds and renovations may not exceed a heat consumption per unit area of 50 W/m^2. The 2009 amendment to EnEV reduced the permitted transmission heat losses by a further 30 per cent. This represents a percentage reduction of around 15 per cent. Consequently, heat consumption per unit area may not exceed 40 W/m^2 to max. 45 W/m^2. A further reduction in heat consumption is planned for 2012. As noted above, a rough calculation using the values above is not a substitute for a precise calculation of heating load in accordance with German standard DIN EN 12831, because this DIN standard takes into account a wide variety of building-specific characteristics, such as solar power generation, etc.

A further 30 per cent reduction in permitted heat loss is planned for 2012.

Where the living area needing heating is known, then the necessary heat output can be roughly calculated as follows:

$$P_H = A \times 45 \ W/m^2$$

where P_H = heat output [W] or [kW]
       A = area of living space to be heated [m^2]

PLEASE NOTE

If a precise calculation of heating load for the building undertaken in accordance with DIN EN 12831 is available, then this calculation takes precedence in determining the required heat pump output.

This standard also sets standard ambient temperatures.

Naturally, it makes sense to use the heat pump both for heating and for generating domestic hot water. In general, we assume 0.25 to 0.35 kW/person is needed for hot water generation alone, so that an average four-person household will require around 1.2 kW. It should also be noted that the higher flow temperatures needed for hot water generation will reduce not only the energy performance ratio, but also the output of the heat pump, which is correspondingly smaller. The following applies:

$$P_{HW} = P_{HWP} \times N$$

where P_{HW} = heat output for hot water generation
       P_{HWP} = heat output for hot water per person, 0.25 kW/person to 0.35 kW/person
       N = number of people

Please note that increased levels of insulation reduces heating requirements, so that greater attention must be paid to the heat output required for hot water generation.

TIP

If domestic hot water is to be generated in an ecologically and economically sensible manner, then this hot water should also be used for running the dishwasher, washing machine, etc. This is two to three times more efficient than using power drawn directly from the grid. In turn, the output needed for hot water generation will be correspondingly higher.

Thus:

$$P_{H'} = P_H + P_{HW}$$

Many utilities offer special tariffs for heat pumps. However, these include blocking periods that need to be taken into account in the calculations. The heat pump must have the necessary 'power reserves' in order to be able to release energy during the blocked periods, even on very cold days with the minimum ambient temperatures as outlined in DIN EN 4701. We use the following factor for these blocking times:

$$S = 24 \text{ h} / (24 \text{ h} - t_{off})$$

where   S = relationship 24 h / access period
         t_{off} = blocking time [h]

The heat output $P_{H''}$ is calculated for a unit area heat output of 45 W/m² as:

$$P_{H''} = P_{H'} \times S$$
$$P_{H''} = (P_H + P_{HW}) \times S$$
$$P_{H''} = (F \times 45 \text{ W/m}^2 + P_{HWP} \times N) \times 24 \text{ h} / (24 \text{ h} - t_{off})$$

where $P_{H''}$ = heat output taking into account blocking periods

Often the standard ambient temperature is interpreted as the minimum temperature. Consequently, there is an assumption that the blocking periods do not need to be considered when planning the heat pump, as they will occur during the day (i.e. when the temperature is warmer).

However, this is incorrect! The standard ambient temperature corresponds to the minimum expected average temperature. For example, the standard ambient temperature for Düsseldorf is −10 °C. However, it is possible for temperatures to fall to −20 °C at night. Daytime temperatures of −2 °C compensate for these night-time lows, resulting in an average temperature of around −10 °C. Operational downtime during blocking periods mean that the building is not being continually fed with the necessary heat. That is why it is very important that utility blocking periods are taken into account.

- *It is important that both the building and the heating system are planned in accordance with the standard ambient temperature.*

The following example should demonstrate this. Our basis is a house with 150 m² living space and four inhabitants (three bedrooms) that needs to be heated:

$$\rightarrow P_H = A \times 45 \text{ W/m}^2$$
$$= 150 \text{ m}^2 \times 45 \text{ W/m}^2$$
$$= 6{,}750 \text{ W}$$
$$= 6.75 \text{ kW}$$

The hot water generation for four people with an average value of 0.3 kW/person:

$$P_{HW} = 0.30 \text{ kW/person} \times 4 \text{ people}$$
$$= 1.2 \text{ kW}$$

$$\rightarrow P_{H'} = P_H + P_{HW}$$
$$= 6.75 \text{ kW} \times 1.2 \text{ kW}$$
$$= 7.95 \text{ kW}$$

This is the minimum necessary heat pump output. Assuming a minimum average ambient temperature, the so-called standard ambient temperature according to DIN EN 12831 Part 1, the heat pump will run for 24 hours. This corresponds to a daily heat output of:

$$Q = 7.95 \text{ kW} \times 24 \text{ h}$$
$$= 190.8 \text{ kWh}$$

However, where t is only 20 hours instead of 24 hours, then the minimum heat output of the heat pump must be designed for

$$P_{H''} = Q / t$$
$$= 154.8 \text{ kWh} / 20 \text{ h}$$
$$= 9.54 \text{ kW}$$
$$= P_{H'} \times S$$
$$= 6.45 \text{ kW} \times 1.7$$
$$= 9.54 \text{ kW}$$

in order to provide the house with the necessary amount of heat, because *the house also cools down during the utility blocking periods.*

For the minimum temperature (e.g. −10 °C), this is the quantity of heat that needs to be fed to the house during the day. Where utility blocking periods mean that power is supplied for fewer than 24 hours, then this must be compensated for by a greater heat output. If the heat fed to the house is regarded as a volume (content) of a rectangle (or several rectangles) with a volume of Q, then the volume is calculated as the product of the two sides, namely heat output P and heating duration t:

$$Q = P \times t$$

where   Q = heat supplied [kWh]
        P = heat output [kW]
        t = heating duration [h]

The following illustrations serve to highlight this:

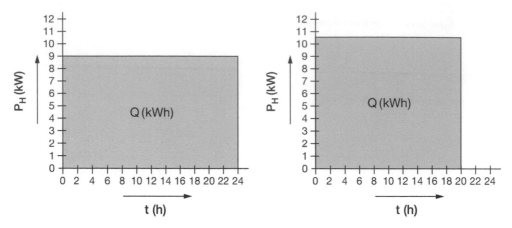

Figure 5.1 Heat output, 24-hour operation   *Figure 5.2* Heat output, 20-hour operation

It makes no difference if the utility blocking times are grouped into a single block or divided into several blocks of time.

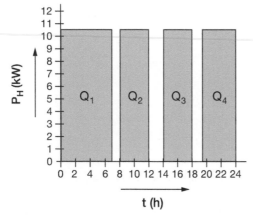

The overall quantity of heat (shaded area(s)) must be the same in each case!

In mathematics this is expressed as an integral over time:

And $Q = Q_1 + Q_2 + Q_3 + Q_4$

where Q_i = quantity of heat between each
                       utility blocking time
             $i = 1–4$

Figure 5.3
Heat output during interrupted 20-hour operation – total of four hours of blocked time

---

TIP

Often the timing of local blocking periods varies significantly so that it is not possible to make a universally valid assertion here. It is even possible that the blocking periods will be lumped together into a single block of time. This will require a correspondingly large buffer tank.

It is often assumed that the utility blocking periods are distributed throughout the day and can therefore be ignored. This is an assumption that should not be made – planning a heat pump properly involves considering the utility blocking periods.

IMPORTANT

Most controllers permit temperatures to be automatically lowered at night (setback time).

This can have disastrous consequences for heat pumps in new and well-insulated houses!

For example, if a night setback from 20 °C to 16 °C or 18 °C is programmed for the period between 22:00 to 06:00, then this is the equivalent of an eight-hour utility blocking period.

New and well-insulated houses will hardly cool down during this period. The controller recognises that the actual temperature is above the specified temperature during the night setback and therefore the heat pump does not turn on during this time. Yet heat is lost through the outer shell of the building and the house cools down. As a consequence, on very cold days the heat pump output may no longer be sufficient to heat the house to the desired temperature.

Now we need to consider the heat source (i.e. groundwater, borehole or ground heat exchangers).

5.2.2 Determining heat pump output for existing buildings

As EnEV does not apply to existing buildings, here we need to carefully determine heat output in accordance with DIN EN 12831.

The first step is to determine whether using a heat pump makes sense at all. The flow temperatures should be as low as possible, and therefore older buildings with convector heaters (often with flow temperatures of 70 °C and above) are often unsuitable for heating with a heat pump. It makes no sense to use a heat pump for flow temperatures over 45 °C/50 °C and above. In such situations, it would be advisable to completely overhaul the heating system in order to achieve the required flow temperatures. For higher flow temperatures, a water-water heat pump has the great advantage of a better level of performance than a brine-water heat pump.

Note:

It is technically possible to build a heat pump that can generate high flow temperatures. However, from the operator's standpoint this usually does not make any economic sense; heat pumps that generate high flow temperatures operate at very high pressure. As a result, the compressors need more power – this significantly reduces heat pump performance and raises operating costs.

However, where low flow temperatures can be achieved then, despite costing more to install in an existing building than a new build, installing a heat pump can be worthwhile because the costs of heating an existing building are significantly greater than those of a new building. This makes the potential savings much larger, while simultaneously providing additional living comfort (underfloor heating for heating and cooling).

Where these preconditions have been met, it is worth calculating the heating demand and getting offers. It may be advisable to instruct an energy consultant or planning office to provide an exact calculation of heating demand.

TIP

Here, too, the advice regarding hot water supply and utility blocking periods in particular applies.

5.2.3 Heat pumps in existing buildings

Due to increasing energy costs, many homeowners want a heat pump in the belief that it will reduce their heating costs. However, this will only be the case where particular framework conditions are in place. Otherwise, nasty surprises may be in store, as frequently discussed on German TV after presenting negative examples of heat pump use.

A colleague at the Oberhausen Chamber of Trade spoke to me about this. "We go to so much trouble to get people interested in heat pumps, and then on TV yesterday we hear a negative presentation about heat pumps. I don't believe it!" Unfortunately, you have to believe it. That was a 2008 TV show. And sadly it is true that many heat pumps have been installed in buildings for which they were not appropriate. There are even manufacturers who claim that their heat pumps can heat to high flow temperatures of 65 °C – even with an air-water heat pump. Although technically possible, in objective, economic and ecological terms it is completely absurd, unless the utilities significantly drop the price for green electricity. This is currently not the case.

To explain this more clearly, let us look at an air-water heat pump with an achievable flow temperature of 65 °C. For an ambient temperature of 0 °C, we have a theoretical energy performance ratio A0W65 of only 2.6. When we look at the oil or gas prices, then it very quickly becomes clear that investing in a heat pump with high flow temperatures makes absolutely no economic sense with electricity prices at their current high levels. If we also take into account the overall performance of the electricity generation process, from the power plant to the end consumer, then such a heat pump would be more harmful to the environment than a gas boiler.

Let us look at the term 'heat pump' again. It pumps heat from a lower temperature level to a higher. The greater the temperature difference, the worse the performance. This is an irrefutable physical law.

The best chance of using a heat pump sensibly in an existing building is where there is underfloor heating, at least on the ground floor. The following chapter examines this situation.

It is less sensible to use a heat pump in an existing building heated exclusively with radiators, something we examine two chapters ahead.

5.2.3.1 Heat pumps in existing buildings with underfloor heating and radiators

As discussed in the previous section, it can make sense to use heat pumps in existing buildings.

There are many older buildings in which the lower floors (usually the ground floor) are heated using underfloor heating, and with radiators on the upper floors. This arrangement is a good precondition for the effective use of a heat pump system so long as the maximum required flow temperature does not exceed 45 °C, max. 50 °C. The larger the share of heating undertaken using underfloor heating, the better the heat pump's coefficient of performance (COP).

Where these preconditions exist, then the following questions need to be answered:

1   What are the maximum flow temperatures needed for the underfloor heating?
2   How much of the heat output is used for underfloor heating?
3   What are the maximum flow temperatures needed for the radiators?
4   How much of the heat output is used for the radiators?
5   Can the heat pump controller regulate two heating circuits – one mixed and one unmixed circuit?

- *When selecting the heat pump, it should be noted that the stated heat output is lower for high flow temperatures than for standard conditions (max. flow temperature 35 °C).*

Two separate buffer tanks are recommended to achieve optimal operation, one for the underfloor heating and one for the radiators. Where space limitations make this impossible, then the buffer tank should at least be filled layer by layer. The following diagram shows a heat pump system with two buffer tanks, with cooling via the underfloor heating.

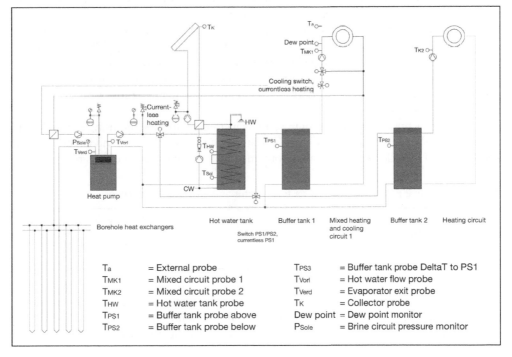

T_a	= External probe		T_{PS3}	= Buffer tank probe DeltaT to PS1
T_{MK1}	= Mixed circuit probe 1		T_{Vorl}	= Hot water flow probe
T_{MK2}	= Mixed circuit probe 2		T_{Verd}	= Evaporator exit probe
T_{HW}	= Hot water tank probe		T_K	= Collector probe
T_{PS1}	= Buffer tank probe above		Dew point	= Dew point monitor
T_{PS2}	= Buffer tank probe below		P_{Sole}	= Brine circuit pressure monitor

Figure 5.4 Brine-water heat pump system with two buffer tanks, a mixed and an unmixed heating circuit – with cooling for the underfloor heating circuit

This diagram shows the appropriate use of a heat pump system for a building with underfloor heating on the ground floor, and radiators on the upper floors. Physics teaches us that, in principle, heat rises. So where the heating system is correctly installed, heat can primarily be fed in via the underfloor heating. This involves low flow temperatures and therefore a high level of performance. The remaining required heat can be fed into the living space via the radiators. This is an example of how a heat pump can be used to heat effectively in an existing building.

Where old radiators are replaced with low-temperature radiators, then this will raise the performance level. Where old radiators are replaced with fan convectors, the overall performance can be raised even further. A second buffer tank may no longer be necessary.

In open-plan buildings, it is very possible that underfloor heating on the ground floor provides almost sufficient warmth to heat both floors, with the radiators on the top floor only needed for supplementary heating. In this case, it may well be advisable to replace the existing radiators with low-temperature radiators, and maybe also to omit the second buffer tank.

Operating with two buffer tanks also makes sense when larger quantities of heat at significantly different temperature levels are required.

5.2.3.2 Heat pumps in existing buildings with radiators

Buildings heated only with radiators are usually unsuited for heat pumps. The building owner must invest in both a heat pump and in converting the heating system to use low flow temperatures before it makes sense to use a heat pump. It often makes no sense to install a heat pump unless such a conversion is undertaken.

The following observations show the importance of exact planning, particularly in existing buildings. The energy consumption data for the period 2007/2008 was collected after oil-fired boilers were exchanged for brine-water and air-water heat pumps with heat outputs of up to 20 kW. They were installed in detached houses and apartment buildings prior to any energy-saving renovations. The required flow temperatures were over 50 °C, and in some cases up to 65 °C.

A variety of brine-water heat pump systems were examined, some using borehole and others ground heat exchangers, as well as air-water heat pump systems for indoor and outdoor installation. The following seasonal performance factors (SPFs) were calculated:

Table 5.1 Seasonal performance factor for brine-water and air-water heat pumps

Brine-water heat pumps with borehole heat exchangers	Brine-water heat pumps with ground heat exchangers	Air-water heat pump systems
3.4	3.2	2.7

The results show that the SPFs are so bad that, in Germany, these heat pumps would not qualify for any funding. As the buildings were previously heated with oil, it must be assumed that the energy costs would be about the same, so that the investment would hardly be worthwhile.

Furthermore, in some of the systems faults were discovered in the hydraulics as well as the control and regulation systems. This included constantly running charging pumps, valves that would not completely close, and missing non-return valves. Therefore, these systems were not running optimally. The electric heating element had also been activated in around 10 per cent of the heat pumps.

In summary, these observations show that in existing buildings, the framework conditions need to be examined in more detail in order to calculate whether a heat pump is worthwhile or not. It may make sense to invest in energy-saving measures to the building and/or the heating system prior to installing a heat pump, including roof and wall insulation, new windows, exchanging radiators for low-temperature radiators or fan convectors, etc.

5.2.3.3 *Save the reputation of the heat pump!*

Back in the 1980s, mistakes were already being made when installing heat pump systems. These were partly the consequence of operators being given incorrect information, partly due to incorrect dimensioning, or simply a lack of knowledge. This was particularly the case for water-water heat pumps, the favoured heat pump system of the time.

Incorrect information and excessive promises were – and are – often given to secure sales.

In terms of dimensioning, the high flow temperatures were often not considered and hydraulic errors were also made. This led to low performance levels and thus low levels of efficiency.

Wells were often dug too deep for water-water heat pumps, and the quality of the well water ignored. Water with a high iron and manganese content clogged up the heat exchangers, and the evaporators broke and/or the discharge wells flooded. Aggressive water led to corrosion in the evaporator and stopped the heat pump working.

As a result of these negative experiences, even today many technicians are still extremely reserved about installing heat pumps, especially water-water heat pumps. However, this fear can be overcome through expert knowledge and (positive) experiences.

On 10 April 2009, I received a fax from a heating engineer with the subject title 'Heat pumps – curse or blessing?' It made clear that a badly planned or even unplanned heat pump installation can also be a curse for the operator, because it generates costs, both when being installed and during subsequent operation. A further curse is its significant environmental impact, because we can hardly assume that such heat pumps are driven with green electricity. In contrast, a good and well-planned heat pump system can be a blessing for everyone, because the operator reduces heating costs and saves money, and reduced CO_2 emissions have less environmental impact.

On 28 April 2008, the German TV show *markt* was extremely critical of heat pumps. They claimed that power utilities, heat pump manufacturers and technicians promise that using a heat pump brings heating cost savings of up to 50 per cent, a promise that, after a few years' operation, is not fulfilled. Their much-lauded cost effectiveness often reveals itself as an excessively optimistic calculation, and technicians installing the heat pumps often use values and data provided by the manufacturer that are not relevant to the heating system being purchased and installed by the customer.

Many customers look for an economic alternative when their old heating systems need replacing. They have often heard of a miracle machine, a heat pump, and believe it to be the solution. And then it is easy for the technician to sell them a heat pump without considering the specific conditions, especially the low flow temperatures that are required. The technician sells a heat pump, often an air-water heat pump, because that is easier.

Added to this, many manufacturers claim their heat pumps can achieve high flow temperatures of up to 65 °C and more. Ideal for renovating older buildings. However, it contravenes the laws of physics to claim that a house can be heated so economically – especially with an air-water heat pump. We first need to invent such a heat pump!

In short, although technically possible, this does not make economic or ecological sense. Nor does it help even if utilities promise much cheaper electricity for operating heat pumps.

Let us look again at the term 'heat pump' from Section 3.1, 'Why is a heat pump called a "heat pump"?'.

Because it 'pumps' heat from a lower temperature – the heat source (air, brine or water) to a higher temperature, namely the flow temperature for heating. The greater the temperature difference between the two, the lower the performance, the level of efficiency and, in the end, its economy, because the heat pump needs more power to 'pump' the heat.

It is also often claimed that a heat pump only makes sense in a low energy house, built according to EnEV. This is completely incorrect. The one and only decisive factor is that a heat pump requires low flow temperatures. This is possible in most existing buildings with underfloor heating, whether a low energy house or not. There is significant potential for making savings, especially in existing buildings, because whether heating costs are reduced from a high energy demand (around 40 per cent) or from a lower energy demand (e.g. new builds according to EnEV, around 50 per cent) makes a large difference. Nor is needing a combination of heating methods (e.g. underfloor heating on the ground floor and radiators upstairs) a reason to decide against a heat pump. With careful planning, a heat pump can certainly bring economic and ecological benefits. Without planning, however, the consequences can be disastrous and lead to nasty surprises. It may make sense to insulate the roof or put in new windows before installing a heat pump.

Figure 5.5 Fan convector

If an existing building does not have surface heating, then other measures are necessary to reduce flow temperatures. Two options are: (1) wall surface heating; or (2) low-temperature fan convectors. Retrospective installation of underfloor heating is a complicated procedure, requiring the existing flooring to be dug up and a new one laid. Problems can occur when this leads to a change in floor height so that doors and stair treads no longer fit.

Furthermore, using an electric heating element as an auxiliary source of heating support should be avoided where possible. Heat pump systems with electric heating elements tend to have smaller dimensions (including the borehole heat exchangers) in order to make them cheaper. But then the operator should not be surprised when, at the end of the year, it turns out that the promised savings have not been made. Cheaper cannot always be good.

Field tests show that incorrectly sized heat pump systems have such bad seasonal performance factors that heating costs can be higher than with the old oil-fired boilers they have replaced. And it is particularly sad if these heat pump systems are being supported with state funding.

Therefore, my appeal to everyone: 'Save the reputation of the heat pump', because even today many heat pumps are being installed where they should not be, or are being badly installed. Where, however, the framework conditions are right, then it makes absolute economic and ecological sense to use a heat pump. However, this requires comprehensive expert knowledge, the relevant experience, and the will to treat customers fairly.

5.2.4 Determining heat pump output for industrial halls

Because of their height, calculations of heat output and heat pump performance for industrial halls cannot be based on only 50 W/m². That would be too little because heat rises, especially in halls. Here, the determining factor is the volume of the industrial hall. We recommend calculating as follows:

The necessary heat output for an industrial hall can be determined using the following formula:

$$Q = V_H \times q \times \Delta T \times f_1 \times f_2 \times f_3 \times f_4$$

where Q = the hall's heat demand [W], [kW]
V_H = hall volume [m³]
q = specific heat demand – see diagram [W/m³K]
ΔT = temperature difference inside/outside [K]
f_1 = correction factor U-value
f_2 – correction factor frequency of doors being opened and hall location
f_3 = correction factor adjacent heated rooms
f_4 = correction factor floor space

First, the volume of the hall is calculated:

$$V_H = L_H \times W_H \times H_H$$

where V_H = hall volume [m³]
L_H = hall length [m]
W_H = hall width [m]
H_H = hall height [m]

The following diagram shows how specific heat demand is calculated:

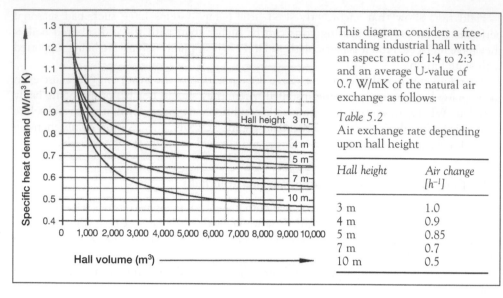

Figure 5.6 Specific heat demand depending upon hall volume

Furthermore, the maximum desired (permitted) temperature difference δT must be agreed with the building owner. The following critical values must be considered:

- Min. ambient temperature according to DIN EN 4701 (–10 °C . . . –18 °C).
- Min. permitted temperatures in accordance with workplace regulations for industrial premises.

The correction factor f_1 for the U-value is as follows:

Table 5.3 Correction factor for U-values

$f_1 = 0.6–0.63$	Very good insulation, average U-value = 0.35 W/m²K, low proportion of surface area taken up by windows and doors: 10%
$f_1 = 0.7$	Normal insulation in accordance with EnEV 2009, average U-value = 0.45 W/m²K, low proportion of surface area taken up by windows and doors: 20%
$f_1 = 1.0$	Insulation in accordance with older thermal protections regulation, average U-value = 0.7 W/m²K, proportion of surface area taken up by windows and doors: 20%
$f_1 = 1.5–2.0$	Insufficient insulation, proportion of surface area taken up by windows and doors: 20%–40%
$f_1 = 2.0–3.0$	No wall insulation/old halls, proportion of surface area taken up by windows and doors: over 40%

The following table applies for correction factor f_2 for door opening frequency and hall location:

Table 5.4 Correction factor for door opening frequency

Normal position	Unprotected position or area subject to heavy winds (coast, hillside)	Door opening frequency
$f_2 = 0.8$	1.0–1.2	Rarely (max. 1 min/h) or doors with automatic closing mechanism
$f_2 = 1.0$	1.3–1.4	Average (max. 5 min/h)
$f_2 = 1.3$–1.5	2.0–2.5	Frequently, no automatic closing mechanism
$f_2 = 2.0$–2.5	3.0–4.0	Doors sited opposite one another, both often simultaneously open (draft)

The following table applies for correction factor f_3 for adjacent heated rooms:

Table 5.5 Correction factor for adjacent heated rooms

	$f_3 = 1.0$		No adjacent heated room
To 2,000 m³	*Over 2,000 m³*	*Height*	
$f_3 = 0.90$ $f_3 = 0.94$	$f_3 = 0.88$ $f_3 = 0.92$	To 5 m Over 5 m	Upper floor, basement heated (no heat loss via floor)
$f_3 = 0.80$ $f_3 = 0.88$	$f_3 = 0.76$ $f_3 = 0.84$	To 5 m Over 5 m	Basement, upper floor heated (no heat loss via ceiling)
$f_3 = 0.97$ $f_3 = 0.95$	$f_3 = 1.00$ $f_3 = 0.97$	To 5 m Over 5 m	Short outside wall, adjacent room heated (no heat loss, short outside wall)
$f_3 = 0.90$ $f_3 = 0.85$	$f_3 = 0.95$ $f_3 = 0.90$	To 5 m Over 5 m	Long outside wall, adjacent room heated (no heat loss, long outside wall)

The following values apply for the correction factor f_4 for floor space:

Table 5.6 Correction factor for floor space

	$f_4 = 1.0$		Normal hall with an aspect ratio of 1:4 or 2:3
To 2,000 m³	*Over 2,000 m³*	*Height*	
$f_4 = 0.96$ $f_4 = 0.94$	$f_4 = 1.00$ $f_4 = 0.97$	To 5 m Over 5 m	Rectangular floor space
$f_4 = 1.06$ $f_4 = 1.10$	$f_4 = 1.04$ $f_4 = 1.08$	To 5 m Over 5 m	Long, narrow hall, aspect ratio around 1:5

Where these values and the correction factors are known, then the heat demand for the hall can be roughly calculated. This provides a rough estimate. Precise calculations in accordance with DIN EN 12831 must be undertaken for special halls, churches, etc., or for a site inspection. This also applies for halls with automatic ventilation.

5.2.5 Sample calculation of the heat output for a detached house in accordance with the German Renewable Energies Heat Act

The Sample family are planning to build a house with the following floor dimensions:

Basement:

- Large basement room   38.75 m²
- Stairwell   7.17 m²
- Laundry room   14.45 m²
- Boiler room   23.08 m²

Ground floor:

- Living room   38.75 m²
- Kitchen   23.08 m²
- Stairwell   17.80 m²
- Hallway and cloakroom   16.99 m²
- Storage room   3.26 m²
- Guest toilet   2.22 m²

Upper floor:

- Bedroom 1   12.51 m²
- Bedroom 2   10.37 m²
- Bedroom 3   10.53 m²
- Stairwell   3.14 m²
- Bathroom   8.83 m²

The customer's request:

The house will be built in accordance with current regulations. The customer does not wish to pay for official heat demand calculations and takes 45 W/m² as the basis for calculations.

In the basement, only the laundry room and large basement room (party room) will be heated.

The customer wishes to equip both basement and upper floor with fan convectors.

There will be underfloor heating on the ground floor.

The first step is to advise the customer that for rooms heated with underfloor heating, the maximum flow temperature should be 35 °C. Reasoning: a heat pump should heat with the lowest possible flow temperatures in order to achieve the maximum levels of performance. The customer agrees.

The next step is to explain to the customer that all rooms in the insulated building must be considered when calculating heating load, because these will be heated slightly by the adjacent rooms, even when not equipped with their own radiators.

Calculating the floor space requiring heating:

- Total basement area   83.45 m^2
- Total ground floor   102.10 m^2
- Total upper floor   45.38 m^2

The total space requiring heating is   230.93 m^2

Using these figures, the minimum required heat output is:

$$P_H = A \times 45 \text{ W/m}^2$$
$$= 231 \text{ m}^2 \times 45 \text{ W/m}^2$$
$$= 10.395 \text{ W}$$
$$= 10.4 \text{ kW}$$

For hot water generation for four people with a higher water requirement (dishwasher, washing machine, etc.):

$$P = 0.35 \text{ kW/person} \times 4 \text{ people}$$
$$= 1.4 \text{ kW}$$

$$\rightarrow \quad P_{H'} - P_H + P_{HW}$$
$$= 10.4 \text{ kW} + 1.4$$
$$= 11.8 \text{ kW}$$

The next task is to ask the local utility for details of their tariffs and any blocking periods. Many German utilities (e.g. RWE) have blocking times spread through the day (e.g. morning 1 h, midday 2 h and evenings 1 h) (i.e. 4 h in total).

Now:

$$S = 24 \text{ h} / (24 \text{ h} - t_{\text{off}})$$
$$= 24 \text{ h} / (24 \text{ h} - 4 \text{ h})$$
$$= 24 \text{ h} / 20 \text{ h}$$
$$= 1.2$$

Thus the heat output $P_{H''}$ is calculated as:

$$P_{H''} = P_{H'} \times S = 11.8 \text{ kW} \times 1.2 = 14.2 \text{ kW}$$

It is advisable to choose the next largest heat pump with output of 15.0 kW.

Table 5.7 Heat output of various types of heat pump

Brine-water heat pump, standard reference: B0/W35				
Type		Heat output	Power output	COP
GeoMax® SW	4,600	4.6	1.1 kW	4.1
GeoMax® SW	5,900	5.9	1.4 kW	4.1
GeoMax® SW	7,000	7.0	1.7 kW	4.2
GeoMax® SW	8,300	8.3	2.0 kW	4.2
GeoMax® SW	10,300	10.3	2.5 kW	4.2
GeoMax® SW	12,400	12.4	3.0 kW	4.2
GeoMax® SW	15,000	(15.0)	3.4 kW	4.4
GeoMax® SW	16,800	16.8	3.8 kW	4.4

Width: 700 mm; height: 900 mm; depth: 450 mm

TIP

Some utilities will have a six-hour blocking time during the day, in a single block. Here, large buffer tanks must be considered and calculated with a factor S = 1.33 (or at least 1.2, as the blocking times are usually during the day).

Note:

The calculations above are based on an ambient temperature of –12 °C for a specific region (in this case, North Rhine Westphalia) and a room temperature for underfloor heating of 20 °C.

However, the homeowner's heating behaviour should also be discussed and taken into account. Some homeowners will want to use underfloor heating to achieve a room temperature of 24 °C, for example. This will require a somewhat higher heat output.

It may be that the homeowner wishes to extend the house at a later stage, adding an extra floor. Therefore, possible future extensions or additions should be taken into account and the heat pump dimensioned accordingly.

These are all points that need to be discussed with the homeowner and heat pump operator.

5.2.6 Determining the heat output for a swimming pool

A swimming pool can be heated very economically and efficiently using a heat pump. This is particularly interesting because a swimming pool can be heated with relatively low flow temperatures.

First, we need to determine whether we are dealing with an open-air pool or an indoor pool.

The heat demand for an open-air pool depends first on the thermal conduction (insulation) beneath, as well as its location and how the pool is used.

Initial values for open-air pools are as follows:

When covered: 50–150 W/m^2
Protected area: 50–200 W/m^2
Partly protected area: 100–300 W/m^2
Unprotected area: 200–500 W/m^2

When heating a swimming pool, a chlorine-resistant stainless steel heat exchanger is necessary to keep the chlorinated water in the pool separate from the system.

TIP

Using a heat exchanger leads to a temperature difference of around 5 K (transmission losses).

The heating demand for indoor pools is significantly less. However, this depends on many factors (e.g. whether a dehumidification system is planned, if a pool cover is used, etc.). A swimming pool cover reduces heat demand by up to 50 per cent, because the heat loss caused by evaporation is significantly reduced. Again, this depends on how often and how long the pool is used.

It is recommended that an energy consultant or swimming pool constructor be consulted for a precise calculation of heat demand.

5.2.7 Heat pumps for commercial use

Energy costs are rising faster than expected. So why not use the energy we have already paid for in a sensible way?

There is a wide range of possible commercial applications for heat pumps, particularly where 'waste' heat can be reused for heat generation. The following questions must be considered when examining the potential use of heat pumps:

- Where, and how much 'waste' heat is available?
- What is the temperature of the available 'waste' heat?
- Where, and how much heat is needed?
- What temperature should the useful heat have?
- How can we connect the 'heat source' (waste heat) with the useful heat (heat demand)?

Here are a few examples:

- Manufacturing industry
  Cooling processes in manufacturing generate large quantities of heat. Rather than allowing this heat to escape, unused, into the atmosphere, we should consider how it could be used effectively. Every manufacturing operation has adjacent office, management or social rooms that need heating; it may even be possible to heat neighbouring buildings via a small community heating network.
- Large office building

Cooling requirements in office buildings are continually increasing as we use ever-greater numbers of electronic devices (computers, photocopiers, telecommunication systems, etc.). Rather than pumping the exhaust heat outside via the air conditioning system, this heat could be used (e.g. to heat hot water for coffee machines, dishwashers, gastronomy areas, etc.).

It might also make sense to use the 'waste' heat to cover the heat demand in other areas of the building.

- Hospitals, hotels, clinics, etc.

  These buildings generate large quantities of exhaust heat through cooling, air conditioning, etc., which could be used very cost-effectively for generating hot water, heating pools, etc. The energy is already there!

- Gastronomy, slaughterhouses

  Large quantities of 'waste' heat are produced in hotels, restaurants, slaughterhouses, etc. as the result of cooling. This heat could be used for heating, generating hot water, etc.

- Agriculture

  Agriculture is also a great source of heat, generated by a variety of sources: from barns, cooling milk, silage heat, etc. This heat can be cleverly used by means of a heat pump to heat pigsties, living rooms, etc. Depending upon animal species and weight, barn temperature and humidity, animals can emit 100 W to 1,000 W of heat:

Table 5.8 Heat emitted by animals

Dairy cows, calves and bulls, young cattle – optimum barn temperature: 16 °C, humidity: 80%							
Animal weight	100 kg	200 kg	300 kg	400 kg	500 kg	600 kg	700 kg	800 kg
Heat emitted per animal	260 W	450 W	620 W	760 W	880 W	980 W	1,050 W	1,100 W
Gilts, sows, boar – optimum barn temperature: 12 °C, humidity: 80%							
Animal weight	40 kg	60 kg	80 kg	100 kg	150 kg	200 kg	250 kg	300 kg
Heat emitted per animal	100 W	135 W	165 W	195 W	265 W	340 W	410 W	485 W

In finding answers to the questions above, economic considerations play a particular role – unfortunately, often too important a role. When it comes to examples of commercial use, the special cheap commercial tariffs offered by the utilities have a decisive impact. However, even these energy costs are rising, making energy recycling increasingly interesting.

The use of any air conditioning or cooling unit should be considered thoroughly. Although air conditioning units are heat pumps, they emit the valuable 'waste' heat, unused, into the atmosphere. Using a heat pump to make use of this emitted heat is a means of helping the environment and your budget.

In general, we should identify 'waste' heat energy that could be sensibly recycled. This is applied environmental protection!

5.2.8 Heat pumps in housing developments and large heat pump systems

Increasingly, 'heat pump housing developments' and large systems with borehole heat exchanger fields are being planned and built. Using heat sources in this way is an excellent way of securing adequate long-term supplies of heat for all buildings.

The following example looks at a 'heat pump housing development' with 45 terraced houses, each with a living area of around 150 m^2, and on a plot of ground of around 250 m^2. Around 7 kW is needed for heating and hot water generation for four people. That represents a cold load of around 5.6 kW. For 45 houses, this represents a total abstraction capacity of 252 kW.

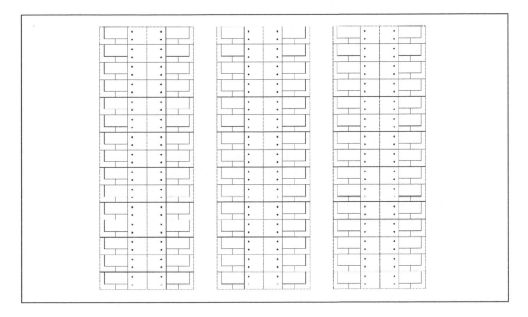

Figure 5.7 Heat pump housing development with 45 terraced houses

Now we will discuss the various types of heat pump that could be used to heat the terraced houses in the housing development.

5.2.8.1 Water-water heat pump systems

Water-water heat pumps draw heat from groundwater. Therefore, it must be assumed that over the long-term, the groundwater can cool significantly. Increased cooling raises the danger of freezing. Therefore, a geological survey that examines groundwater movements, volumes and water quality is recommended.

If every house has its own intake and discharge well, then the danger of the groundwater cooling near the surface is very great. Consequently, over time, the danger of heat pumps on the housing development freezing increases.

It may make sense to install a centralised water supply (i.e. a large intake well and a large discharge well), which supplies each house on the entire housing development via a ring pipe. The local heating system in Dorsten-Wulfen, Germany, is a good example of such a layout.

The heat pump housing development shown above has an overall water demand of around 40 m³/h to 50 m³/h. Consequently, the ground must have a sufficiently large run-out capacity and, in particular, a large absorption capacity. It may be worth asking the local authorities if they would be prepared to provide cost-efficient drainage for such an environmentally friendly project. Then all the heat pumps could be fed via a single intake well and, rather than discharge wells, a single water pipe could feed the discharged water into a stream, a river or a lake. This would completely avoid the long-term cooling of the aquifer, thereby increasing the security of each individual heat pump system.

5.2.8.2 Brine-water heat pump systems

Generally, in densely populated areas, individual land plots are too small to install ground heat exchangers, leaving borehole heat exchangers as the only option. For larger heat pump systems or 'heat pump housing developments', it is vital to ensure that heat is not drawn only from particular points in the ground, but that extraction is spread over a large area and a large volume. Over a period of several years, the extraction capacity can fall significantly and therefore the geological characteristics must be also considered. Reductions in abstraction capacity of 20 per cent and more are entirely realistic. A Geothermal Response Test is essential for larger heat pump housing developments.

The borehole heat exchangers extract the geothermal energy via thermal conduction. If a large number of exchangers are installed within a set area, then the outer exchangers will conduct sufficient quantities of geothermal energy, whereas, over time, the inner ones will not. The following diagram demonstrates this clearly.

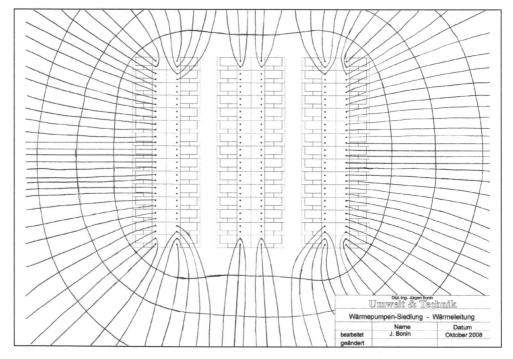

Figure 5.8 Heat pump housing development with borehole heat exchangers – isothermal lines and field lines for thermal conduction – 45 terraced houses with borehole heat exchangers

If a 'heat pump housing development' is planned using this layout, then we have to assume that over the long-term, the heat pumps connected to the inner exchangers will be subject to low pressure errors because they will no longer be fed with sufficient geothermal energy. The temperature level sinks below that necessary for the heat pump to operate properly. In extreme cases, these borehole heat exchangers can even become clogged.

The lines leading to the borehole heat exchangers show the thermal conduction (mathematical term: field lines), and the lines circling the exchangers indicate the fall in temperature (mathematical term: lines of equal potential). Using the relevant programmes, the path of these lines can be calculated with greater precision and in three dimensions.

The borehole heat exchangers shown in the example above should be distributed better, not in rows and columns as here. More exchangers should be located in the outer areas, and those in the centre should be placed further apart, ensuring that each borehole heat exchanger has roughly the same abstraction capacity. It may also be sensible to drill the inner exchangers more deeply than the outer ones. Again, this will depend on the local geology.

In this instance, cooling is doubly relevant: not only does it offer the luxury of cooling on hot days; it also helps the ground to thermally regenerate. The solution is shown as follows:

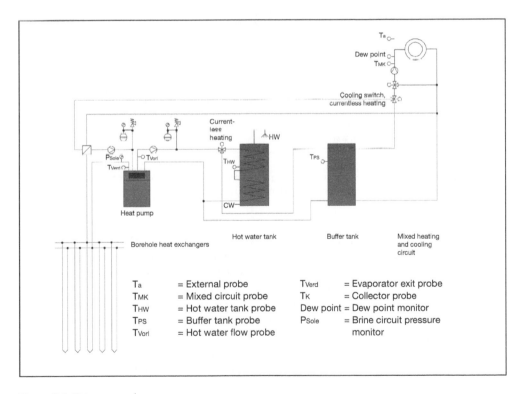

Figure 5.9 Brine-water heat pump system

The heat extracted from the house during cooling is returned to the ground for later use during heating periods. For larger heat pump housing developments, solar power can also be used to regenerate the ground more effectively:

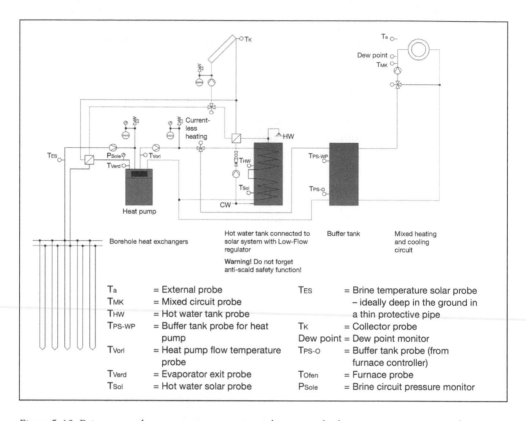

Figure 5.10 Brine-water heat pump system using solar power for hot water generation and to regenerate the borehole heat exchangers

Using this heat pump system, the solar power can be almost 100 per cent utilised in summer. This is not possible for solar systems with auxiliary heating, because the controller turns off in summer when the tank is full. As the ground has a much greater absorption capacity, it can use almost 100 per cent of the solar power: in summer, heat is pumped into the ground, warming it up. This stored heat increases the brine temperature, making the heat pump work much more efficiently. So the solar power generated in summer is then used later for heating.

A larger buffer tank with solar register can also be used, which the solar system can charge directly during the transition periods. In summer, only once this buffer tank is full is heat then returned to the ground.

A combination of solar power generation and cooling is possible if a reversible heat pump is used. The heat extracted during cooling plays an additional role in regenerating the ground's heat reserves. This is a good way of combining luxury and heat recovery.

This description shows clearly that even with a heat pump, it is important to use the available energy economically. With heating and cooling, and the use of solar power, energy is not only drawn from the ground, but also partially returned. This makes the long-term, secure use of geothermal energy possible if heat pump housing developments and large systems are planned properly. Geothermal Response Tests (GRTs) are a necessary part of correct planning. These calculate the thermal resistance and thermal conductivity of the ground and, in turn, these calculations are used to assess the long-term impact and to dimension the borehole heat exchangers properly.

The geological characteristics play a significant role in determining how the heat is fed to the borehole (or ground) heat exchangers. For borehole heat exchangers, heat can only be transmitted by thermal conductivity (e.g. worst-case scenario – solid rock). The best-case scenario is for an exchanger to be inserted into water-bearing sand or gravel where it can be 'bathed' in water. 'Bathing' does not, however, imply high flow rates; the groundwater flow rates are usually only a few millimetres or less per day.

Precise planning must be carried out by a planning office specialising in geothermal energy.

The direct use of solar power, generated using a solar system, can be used to counteract the ground cooling. It can be used to generate hot water in summer, directing excess energy back into the ground. Consequently, here, the solar power can be almost 100 per cent utilised – for generating hot water and supporting heating. This also significantly improves the seasonal performance factor. However, care must be taken that the brine pipes and borehole heat exchangers do not overheat. Therefore, thermally resistant materials must be selected for use in these systems.

Here, vacuum tube collectors are particularly suitable as they have a much better performance during the transitional periods. However, the maximum temperature must be limited in order to protect the heat exchangers and the brine circuit manifold.

5.2.8.3 Air-water heat pump systems

As is well known, air-water heat pump systems do not involve the costly installation of borehole or ground heat exchangers to tap the heat source(s). However, one disadvantage associated with many air-water heat pumps is the high noise levels. When several heat pumps operate, creating a 'heat pump choir', then this can lead to considerable noise pollution. And when several heat pumps of the same make operate simultaneously, then this can lead to the so-called beat effect; these are humming, fluctuating noises, generated when the noise frequency of the heat pumps differs only minimally. This limits the use of these heat pumps. Air-water heat pumps can be a good and sensible alternative in commercial premises, where noise emissions are not disruptive. However, in private residential areas, only quieter air-water heat pumps are recommended in order to avoid disturbances and disputes. Here, compact heat pumps (evaporator and heat pump in housings) are more suitable, standing as a unit in the basement.

5.2.8.4 Final considerations on heat pumps in residential areas and large heat pump systems

When planning a so-called heat pump housing development (e.g. a new residential area in which the housing stock is heated by heat pumps), local authorities need to plan carefully in advance in order to avoid unpleasant side effects at a later stage. Using a variety of technologies may also be a solution.

5.2.9 Heat pumps for heavily insulated houses and passive houses

There are several terms for housing insulation in Germany: 'KfW houses' or '6 litre' or '3 litre' houses. 'KfW houses' is a reference to the various levels of funding offered by Germany's Reconstruction and Loan Corporation, the *Kreditanstalt für Wiederaufbau* (KfW) (in accordance with EnEV 2004).

The annual primary energy demand of a KfW 60 house is around 60 kWh/m²a. This corresponds with the KfW efficiency house 70 (in accordance with EnEV 2007).

The annual heating demand for a KfW 40 house is a further 30 per cent lower than for a KfW 60 house, and the annual heating demand may not exceed 40 kWh/m²a. This corresponds with the KfW efficiency house 55 (in accordance with EnEV 2007).

There are also other definitions relating to KfW efficiency houses, right through to passive houses, but we will not go into more detail here.

For so-called 6 and 3 litre houses, the primary energy demand is the equivalent to only 6 or 3 litres of fuel oil, around 60 kWh/m²a or 30 kWh/m²a, including all heat pump system components such as pumps, controllers, etc.

Passive houses are buildings in which thermal comfort is achieved solely through the reheating or cooling of the fresh or supply air, in accordance with German air quality standard DIN 1946, and without using circulating air. As the term suggests, passive heat sources are usually sufficient, and include solar radiation, radiation of heat from humans and other internal sources. The area-specific heating demand may not exceed 15 kWh/m²a. It is important too that hot water generation is adequate. Here, combined heat recovery systems with a small, integrated heat pump for generating hot water can be used. The exhaust air from the ventilation equipment serves as the heat source.

The details given above, together with the SPF, enable a rough calculation of the general permitted heating load. This is, however, not a substitute for a precise calculation of heating load according to DIN EN 12831.

These houses have such a low heating demand that oil-fired heating is completely impossible for the following reasons:

• The heat output of any oil-fired boiler would be far too large.
• Oil heating with a tank (as well as a buffer tank) and space for the tanks would be more expensive than any heat pump.

Nor is gas heating with a modulating gas boiler suitable for these houses.

Here, a heat pump is an ideal and exceptionally economic method of heating, especially as a heat pump system offers the additional luxury of cooling, something particularly welcome in summer in these extremely well-insulated houses.

5.3 Planning a water-water heat pump system

Careful planning is vital to ensure that a heat pump system can operate optimally. This is valid both for determining the correct heat pump output, and for the correct dimensioning of the hydraulics for the entire heat pump system, including the connected heating system.

In a water-water heat pump, water is pumped to the heat pump from the groundwater aquifer. Heat is extracted from this water by cooling it down. The cooled water is then returned to the groundwater via the discharge well. The most efficient heat pumps are indisputably water-water heat pumps. The following basic conditions must be met to operate water-water heat pumps safely over the long term:

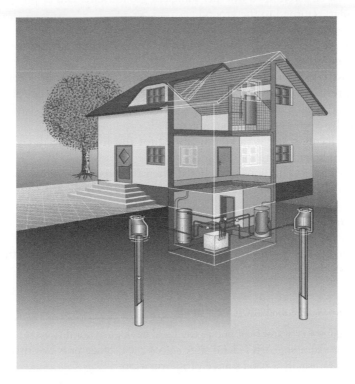

Figure 5.11 Water-water heat pump system

IMPORTANT

- The water must be free of iron and manganese. Other criteria relating to water quality (water corrosiveness) must also be met. See notes on corrosion resistance limits. Where these conditions are not met, then a water-water heat pump is not recommended.
- The discharge well must be able to absorb the quantities of returning water. This is difficult in clay and impermeable ground, or where the water table is very high.

  The quantity of returning water depends on the heat pump and its output. Where these preconditions are not met, then a water-water heat pump is not recommended.

- Where the level of the water table is low, the return line to the discharge well must be installed at a level as deep as that of the intake line. Here, the principle of communicating vessels applies. The water falling back into the discharge well ensures pressure equalisation.

Procedure:

1   First, it must be determined whether sufficient groundwater can be pumped out of and, in particular, returned to the ground.
2   The well water must be examined for iron or manganese content. Where the acceptable limits are exceeded (Fe < 0.2 mg/l, Mn < 0.05 mg/l), then a water-water heat pump is not recommended.
3   The corrosiveness of the water must be determined. Where water is corrosive, then a corrosion-resistant (brazed) heat exchanger must be used, or there must be system separation. System separation involves additional transmission losses and greater energy input for the extra circulation pump required.
4   A suitable location needs to be established for the intake and discharge wells. The relevant minimum separation distances must be observed.
5   Where the above conditions are fulfilled, then the pipe dimensions must be planned.
6   The size of the underwater pump is determined based on the total pressure losses.
     Warning!
     Do not forget that the water level in the well sinks during pumping – and as the well gets older, sinking increases.
7   Do not forget that the necessary permissions must be obtained from the local water authorities.

5.3.1 Hydrological and geological conditions

Before planning a water-water heat pump, the local geological conditions must be examined. In Germany, for example, each federal state's geological office provides this service. Local well engineers are usually a good source of information.

Note:
If well engineers offer both wells and borehole heat exchangers, is it possible that they will favour borehole heat exchangers because these are more profitable to install.

When examining the hydrological conditions, it is not the target extraction rate that is important, but rather the absorption potential of the discharge well. This, in turn, depends upon the following factors:

1   The water yield and absorption capacity of the ground. The greater the yield, the better the ground's absorption capacity. But it can never be assumed that the ground will absorb as much water as it yields.
2   The absorption capacity also depends on the level of the water table. Where the water table is high and the absorption capacity too low, then the discharge well may not take up the returning water.

Where appropriate, a closed well head can prevent water escaping. Care must be taken that the returning water does not flood on to the ground next to the well.

If the intake well yields sufficient water but the discharge well is unable to sufficiently reabsorb the returning water, then it should be determined whether the water can be diverted elsewhere, for example to a ditch, nearby lake, river or similar. This will need to be done in agreement with the relevant authorities.

IMPORTANT

As the water must be almost completely free of iron and manganese, the well should only be as deep as necessary – and not as deep as possible!

The deeper the well, the greater the probability that the water will contain iron or manganese, making it unsuitable for operating a heat pump. This is particularly the case when drilling down through layers of clay.

TIP

With regard to water quality, it is often helpful to have neighbouring wells (ideally, irrigation wells) examined for their iron and manganese content. It can often be assumed that wells drilled close by and to the same depth will have a comparable water quality.

Where all the hydrological conditions have been fulfilled – yield, absorption capacity, lack of iron or manganese – then a water-water heat pump is recommended.

• A further advantage is that the discharge well can also be used for garden irrigation.

Where, despite examinations, doubt remains, one option is to drill the discharge well and analyse its water. If the water is not suitable for a heat pump, then at least the owner now has a well for irrigating the garden.

After the well has been drilled, the water must be subjected to comprehensive analysis to determine its exact composition.

5.3.2 Water quality

The first step is to determine if a water-water heat pump is appropriate. The relevant limit values should be obtained from the heat pump manufacturer/supplier. They should also be asked if they can supply the heat pump with a brazed evaporator (heat exchanger).

An example of the relevant limit values is given in Table 5.9.

TIP

This table is only an example and does not apply for all types of heat pump. Request the relevant data from each manufacturer.

Table 5.9 Limit values for use of a heat exchanger (dependent on manufacturer)

Water content + specific values	Plate heat exchanger, copper brazed	Plate heat exchanger, brazed
pH value	7–9 (considering SI index)	6–10
Saturation index SI (delta pH value)	−0.2 < 0 < +0.2	Not specified
Total hardness	6–15 °dH	6–15 °dH
Permeability	10–500 μS/cm	Not specified
Suspended solids	< 30 mg/l	< 30 mg/l
Chloride (Cl⁻)	< 500 mg/l	< 500 mg/l
Free chlorine	< 0.5 mg/l	< 0.5 mg/l
Hydrogen sulphide (H_2S)	< 0.05 mg/l	Not specified
Ammonia (NH_3/N_4)	< 2 mg/l	Not specified
Sulphate (SO_4)	< 100 mg/l	< 300 mg/l
Hydrogen carbonate (HCO_3^-)	< 300 mg/l	Not specified
Hydrogen carbonate/sulphate	> 0.1 mg/l	Not specified
Sulphide (SO_3^-)	< 1 mg/l	< 5 mg/l
Nitrate (NO_3)	< 100 mg/l	Not specified
Nitrite (NO_2)	< 0.1 mg/l	Not specified
Iron (Fe)	< 0.1 mg/l*	< 0.1 mg/l*
Manganese (Mn)	< 0.1 mg/l*	< 0.1 mg/l*
Free, aggressive carbonic acid (H_2CO_3)	< 20 mg/l	Not specified

* It makes no sense to install a water-water heat pump where these limits are exceeded. It must be assumed that the well will become clogged over time. The evaporator can also clog up and, in time, corrode – electrolytic corrosion (stainless steel and iron).

Often a water-water heat pump requires a brazed heat exchanger. Where the well water is corrosive a heat exchanger for system separation is often recommended as an alternative to brazed evaporators. However, this does involve the following disadvantages:

1   An additional circulation pump is needed for the intermediate circuit, including safety groups and expansion tank.
    This is shown schematically in Figure 5.12.
2   Antifreeze must be added to the intermediate circuit to protect the evaporator and prevent it from freezing.
3   The heat exchanger used for system separation must be regularly checked so that it can be exchanged quickly if there is a defect, and especially if it corrodes. A brazed heat exchanger is highly recommended as a means of avoiding corrosion, especially for aggressive water. So consider including a brazed heat exchanger in the heat pump right from the start.
4   The overall performance level is reduced because of the energy needed to drive the extra circulation pump and the lower temperatures at the evaporator (transmission losses).

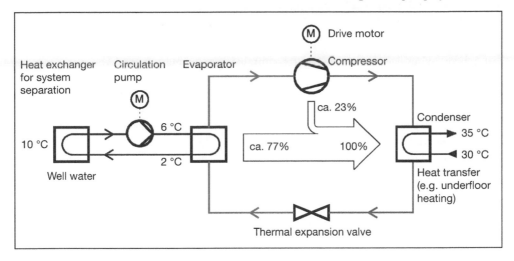

Figure 5.12 Water-water heat pump with system separation

5.3.3 Intake and discharge wells

The intake and discharge wells are drilled by a well engineer. They must provide the required extraction capacity, and especially the absorption capacity, over the long-term and under constant operation. The discharge well often has a longer filter route than the intake well. The underwater pumps that are used will determine the extraction and absorption capacities.

TIP

You should not only consider the minimum flow through the heat pump, but also the extraction capacity of the next largest underwater pump + reserve.

The direction of flow must be taken into account when positioning the wells. The discharge well should be positioned so that the cooled return water being fed into it flows away from the intake well.

Furthermore, there must be a minimum distance between the wells to avoid a thermal short circuit. This distance will depend primarily on the extraction capacity and thus the heat output.

The well engineer is responsible for dimensioning the wells. Here again, a few points to be considered:

- k_f value (permeability) – the lower the permeability, the longer the filter route needs to be; or a larger drilling diameter (not pipe diameter) can be selected.
- Groundwater levels, so that the discharge well does not overflow.
- Separating the various aquifers with clay barriers.
- Closing off the wells with special well heads.
- Well rooms made from concrete rings to provide permanent access to the wells.

5.3.4 A water-water heat pump where the water contains iron or manganese

For heat pump systems larger than 20 kW and above, it may make sense to use a water-water heat pump even where the water contains iron or manganese. The water can be treated using subterranean water preparation systems. The extracted water no longer needs to be treated once it has left the well, and is returned, unchanged, into the groundwater.

This combination is a worthwhile alternative to borehole heat exchangers, especially for larger heat pump systems. It is also a solution for customers drawing their own water supplies, allowing them to simultaneously treat their drinking water.

This procedure requires two intake wells and one discharge well. The two intake wells are used alternately (i.e. when one is in operation, the other is being enriched with oxygen). After a certain volume of water has been extracted, the system switches from one well to the other. Oxygen enrichment functions as follows: a portion of the water being extracted is very strongly enriched with oxygen drawn from the ambient air before being fed into the intake well that is not currently in operation. This enriches the groundwater reservoir with oxygen, activating the natural treatment process in the soil.

This enables the heat pump to achieve a greater level of efficiency at the same, or only slightly higher, cost.

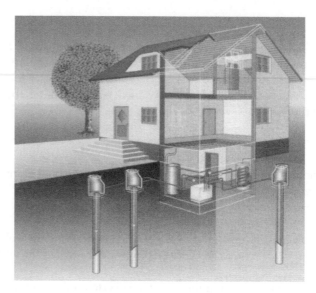

Figure 5.13 Water-water heat pump system with subterranean treatment for iron and manganese

5.3.5 Example of water-water heat pump planning

The heat output for a detached house was calculated in Section 5.2.5, 'Sample calculation of the heat output for a detached house in accordance with the German Renewable Energies Heat Act':

$$P_{H'} = P_H + P_{HW}$$
$$= 10.4 \text{ kW} + 1.4$$
$$= 11.8 \text{ kW}$$

Table 5.10 From a GeoMax® heat pump brochure

Water-water heat pump, reference standard: W10/W35

Model		Heat output	Electrical output	COP
GeoMax® WW	6,400	6.4 kW	1.1 kW	5.8
GeoMax® WW	8,300	8.3 kW	1.4 kW	5.9
GeoMax® WW	9,800	9.8 kW	1.7 kW	5.9
GeoMax® WW	11,600	11.6 kW	2.0 kW	5.9
GeoMax® WW	14,200	14.2 kW	2.4 kW	5.9
GeoMax® WW	17,500	17.5 kW	3.0 kW	5.9
GeoMax® WW	20,900	20.9 kW	3.5 kW	6.0
GeoMax® WW	23,500	23.5 kW	3.9 kW	6.1

Width: 700 mm; height: 900 mm; depth: 450 mm

If the customer does not wish to take advantage of special tariffs offered by the utility (or these are not offered), then the water-water heat pump with an output of 11.6 kW can be used. Assuming a blocking period of four hours, the heat pump with a heat output of 14.2 kW should be used.

The heat pump system can now be installed as follows:

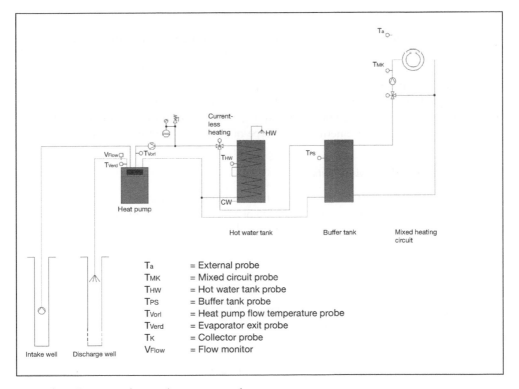

Ta = External probe
TMK = Mixed circuit probe
THW = Hot water tank probe
TPS = Buffer tank probe
TVorl = Heat pump flow temperature probe
TVerd = Evaporator exit probe
TK = Collector probe
VFlow = Flow monitor

Figure 5.14 Diagram of a simple water-water heat pump system

5.4 Planning a brine-water heat pump system

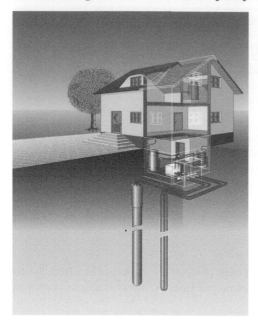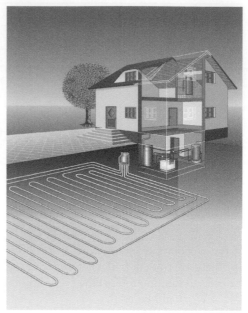

Figure 5.15 Brine-water heat pump system, left with borehole heat exchangers, and right with ground heat exchangers

Brine-water heat pumps work in a closed system, ensuring safe operation over the long-term and making the system almost independent of the groundwater. The heat source is tapped by means of borehole or ground heat exchangers. The heat source is dimensioned in accordance with the calculated cooling capacity.

Permission from the relevant local water authorities is still required for a brine-water heat pump system. Depending upon location (e.g. within a drinking water extraction area), there may be specific requirements for the type of antifreeze used in the brine.

5.4.1 Calculating the cooling capacity

The cooling capacity must be known in order to plan the heat source (e.g. borehole or ground heat exchangers). The cooling capacity of a heat pump is calculated as the difference between the heat pump output and the electrical power consumed by the heat pump:

$$P_C = P_{H'} - P_E$$

where P_C = cooling capacity [kW]
  $P_{H''}$ = heat output of the heat pump [kW]
  P_E = power consumption [kW]

where $P_E = P_{H''} / COP$
  → $P_C = P_{H''} - P_{H''} / COP = P_{H''} \times (1 - 1 / COP)$

The heat output of the heat pump is used to calculate the required cooling capacity and abstraction capacity of the borehole heat exchangers. This also applies for heat pump systems operating with utility blocking periods. Here too, the heat output of the heat pump is taken as the basis for calculations. The laws of physics would suggest that it would make sense to start with the heat output required by the building, including hot water generation, but this is not standard engineering practice. For longer blocking periods and where the building has low thermic inertia, it might be necessary to install a larger buffer tank.

For the purposes of simplification, the following calculation has been proven in practice:

TIP

In a first step, the abstraction capacity P_C can also be simply calculated as:

$$P_C = P_{HHP} \times 0.8$$

where P_C = cooling capacity = abstraction capacity from the environment
               P_{HHP} = heat output of the heat pump

The COP is usually (under standard operating conditions) better than 4, so that the factor 0.8 used above (for COP = 4, factor = 0.75) already accounts for necessary reserves.

5.4.2 Borehole heat exchangers

There are various types of borehole heat exchangers. We differentiate between simple U-exchangers, double U-exchangers, pipe-in-pipe exchangers and direct evaporation exchangers. Simple U-exchangers consist of two PE pipes, one for falling and one for rising brine.

Double U-exchangers are effectively two simple U-exchangers. As they consist of four PE pipes, they have double the surface area over which to absorb geothermal energy. A further important advantage over simple U-exchangers is that the pressure loss is halved.

A pipe-in-pipe borehole heat exchanger consists of an external pipe and an internal pipe.

The number of borehole heat exchangers required depends on the geogenic conditions. The standard lengths are between 50 m and 100 m. The following rules of pressure loss apply:

- The greater the length of a borehole heat exchanger, the greater the pressure loss.
- The smaller the cross-section of the PE pipe, the greater the pressure loss.
- For a set overall length of pipe, the more individual borehole heat exchangers are used, the lower the pressure loss.

The number of borehole heat exchangers must first be agreed with the well engineer before selecting the brine circuit manifold.

We distinguish between two types of exchangers:

- double U-exchangers; and
- pipe-in-pipe exchangers.

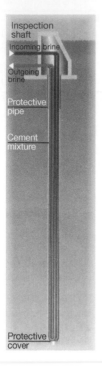

Figure 5.16
Borehole heat exchangers, left double
U-exchangers, right pipe-in-pipe exchangers

Figure 5.17
Installing borehole heat exchangers

Double U-exchangers are normally used because they have several advantages over the pipe-in-pipe exchangers:

- easier to install and therefore more economical; and
- better absorption of geothermal heat due to their larger surface area.

Direct evaporators are another option. These are more efficient because of the direct subterranean evaporation. However, this is counteracted by the relatively high costs of completely gas-sealed ground and borehole heat exchangers. Here, PE pipes are unsuitable, and there are also higher costs because of the significantly larger quantities of antifreeze they need.

5.4.2.1 Planning a borehole heat exchanger

It is particularly important that borehole heat exchangers are sufficiently generously dimensioned. If a heat pump has a cooling capacity of P_C, then the specific abstraction capacity $P_{BHEspec}$ of a borehole heat exchanger can be calculated using the total length L:

$$P_{BHEspec} = P_C / L$$

where $P_{BHEspec}$ = specific abstraction capacity of borehole heat exchanger [W/m]
            P_C = cooling capacity [kW]
            L = length of the borehole heat exchanger [m]

When a heat pump operates, it extracts the cooling capacity of the heat pump from the heat source. Consequently, the heat source, here the ground, is cooled. If a borehole heat exchanger is too small, then too much heat is extracted from the ground and can lead to the ground freezing. When the heat pump is no longer supplied with sufficient geothermal energy, then the low pressure switch kicks in and shuts down the heat pump. It should also be noted that when brine temperatures are too low, the heat pump operates less efficiently.

The following guide values apply for the abstraction capacity according to the German guideline VDI 4650 and for the cooling capacity for borehole heat exchangers (cooling):

Table 5.11 Specific abstraction capacity for borehole heat exchangers

Soil structure	Specific abstraction capacity		Cooling capacity
	1,800 h/a	2,400 h/a	
Dry sand, gravel	< 25 W/m	< 20 W/m	< 10 W/m
Loose dry rock	20–25 W/m	15–20 W/m	9–12 W/m
Dry clay, loam	20–30 W/m	20–25 W/m	12–19 W/m
Wet clay, loam	35–50 W/m	30–40 W/m	18–25 W/m
Fine rock with low thermal conduction	40–45 W/m	35–40 W/m	21–28 W/m
Loose water-bearing rock	50–55 W/m	45–50 W/m	28–31 W/m
Calcareous sandstone	55–70 W/m	45–60 W/m	28–37 W/m
Water-bearing sand, gravel	65–80 W/m	55–75 W/m	34–40 W/m
Sandstone	60–70 W/m	55–65 W/m	34–40 W/m
Acidic magnetite (e.g. granite)	65–80 W/m	55–70 W/m	37–43 W/m
Alkaline magnetite (e.g. basalt)	40–65 W/m	35–55 W/m	20–38 W/m
Gneiss	70–85 W/m	60–70 W/m	35–40 W/m

The specific abstraction capacity values can fluctuate significantly depending upon the local geological characteristics, such as rock formation, stratification, weathering, etc.

When planning the borehole heat exchangers, the heat pump's annual operating hours must also be considered when selecting the specific abstraction capacity. For example, if the output of a heat pump is only just sufficient, then the number of annual operating hours may be significantly greater than shown above. For further details on dimensioning borehole heat exchangers, please refer to German guideline VDI 4640.

Thus the required length L of the borehole heat exchangers can be calculated as follows:

$$L = P_C / P_{BHEspec}$$

IMPORTANT

When calculating exchanger lengths, it should be noted that the abstraction capacity of different ground layers can vary significantly.

Usually the total length of the exchangers is greater than 100 m. Therefore, several borehole heat exchangers are drilled and connected to a circuit manifold. The circuit manifold, in turn, is connected to the heat pump.

In most cases, double U-exchangers are used. These usually take the form of four parallel PE pipes running vertically, DN 20 mm (¾″). For double exchangers the following applies:

$$N_M = 2\ N_{BHE}$$

where N_{BHE} = number of borehole heat exchangers
N_M = number of connections on the manifold

Pipe-in-pipe borehole heat exchangers are seldom used.

There are also other ways of extracting geothermal energy, for example with so-called 'energy posts' that primarily function as foundations. They can contain PE pipes through which geothermal energy is extracted for use in a heat pump. Thus they fulfil a double function, both as a foundation and a heat source. Energy posts are usually only used in larger constructions.

Furthermore, care must be taken that sufficient brine flows through the borehole heat exchangers and that the volume of brine is evenly distributed between the exchangers. Therefore, a circuit manifold with flow plate is recommended. Often no flow plate is offered or installed for reasons of cost. If this is the case, then note the following:

Figure 5.18
Energy posts installed as foundations

Figure 5.19 Energy post prior to installation

TIP

Where the circuit manifold has no flow plate (often for reasons of cost), then it is essential that all PE exchanger loops are the same length, and installed and connected according to Tichelmann.[1]

However, this does not enable control of the flow volumes in each exchanger.

Figure 5.20 Brine circuit manifold with flow plate

According to VDI 4640, the borehole heat exchangers must be connected to a circuit manifold with shut-off device. This makes regulation easier, and if there is a fault then the affected borehole heat exchanger can be closed off.

Borehole heat exchangers form their own, separate circuit, independent of groundwater, a so-called brine cycle. This ensures safe long-term operation. Clogging with iron or manganese is impossible.

In addition, borehole heat exchangers, energy posts and ground heat exchangers should be regarded as real estate. They are installed once and are permanently part of the house. They can supply energy for a whole lifetime and are therefore an investment for life.

As thermal energy is always drawn from the same source (i.e. the ground surrounding the borehole heat exchanger), then over time this ground cools. To avoid excessive local cooling, the borehole heat exchangers must be of a sufficient size. If they are too small, there is the danger of the ground freezing. This results in the heat pump being shut down due to a lack of pressure.

1 Albert Tichelmann, heating engineer, 1861–1926.

When the heat pump is used for cooling in summer, the heat extracted from the building is fed back into the ground where it can be used later for heating. This increases the performance of the heat pump system. The following illustration shows the clear advantage of borehole heat exchangers over ground heat exchangers.

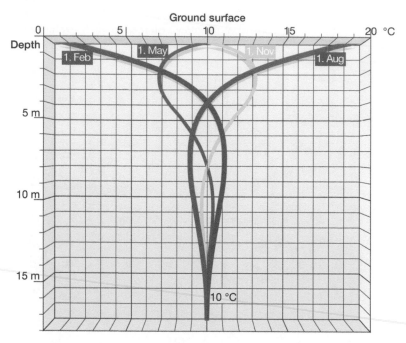

Figure 5.21 Temperature curves in the ground

From a depth of 10 m onwards, one can assume a constant temperature of 10 °C – where no heat pump is operating and depending on the local geogenic characteristics of the ground. A heat pump draws heat out of the ground and the temperature sinks to around 0 °C. 0 °C is the standard reference value according to German standard DIN EN 14511-2.

What happens if the neighbour also heats their house with a heat pump? Does this lead to excess cooling of the ground? What happens in the ground?

Let us look at Figures 5.21 and 5.22.

These diagrams of model calculations show that local ground cooling is spatially relatively limited. Mathematically, we speak of a 'sink', in which the surrounding temperature of the ground sinks.

However, this does not mean that we do not need to consider interactions between the exchangers for larger housing developments heated with heat pumps. Over time, the abstraction capacity can fall significantly for large area or large volume heat extraction.

Long-term temperature projections can be determined using the GEOprog© programme.

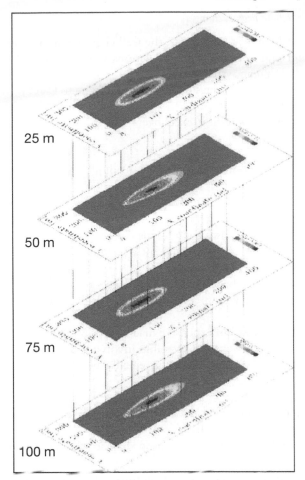

Figure 5.22 Temperatures around borehole heat exchangers – by depth

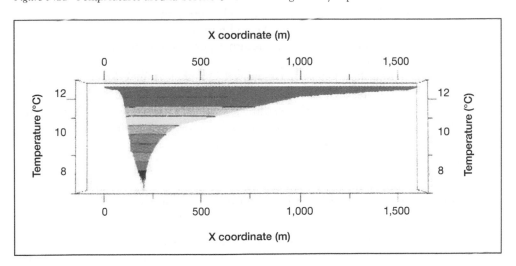

Figure 5.23 Temperatures around borehole heat exchangers – horizontal spread

5.4.2.2 Installing borehole heat exchangers

When installing borehole heat exchangers, the well engineer must consider and adhere to various groundwater and drinking water regulations (see Section 5.9, 'Applications and authorisations for heat pump systems').

In Germany, the well engineer must be certified according to DVGW W120 before being permitted to undertake drilling.

The installation of borehole heat exchangers and the work of the well engineer can involve the following:

1   Preparing the construction area, including loading, transporting and preparing the drilling equipment.
2   Drilling using the correct drilling technique for the class of rock and ground.
3   Installing the borehole heat exchangers and compacting the layers of earth with the relevant clay and cement suspensions. This compression is vital in order to keep the individual groundwater aquifers permanently separate from one another.
4   If applicable, installing a protective pipe.
5   Applying pressure to the borehole heat exchangers to check for any leakages.
6   Drawing up the documentation: a bore log and a drilling profile.

The well engineer has invested in expensive machinery and therefore it is naturally in his interest to undertake as much work as possible involving these machines, especially the drilling equipment. The following tasks can equally be carried out either by the well engineer, or by other specialists:

7   Digging the trenches for laying the connecting pipes to connect the borehole heat exchangers to the distribution shaft.
8   Delivery and installation of the distribution shaft in which the brine circuit manifold will be housed.
9   Installing the brine circuit manifold and connecting the borehole heat exchangers to the brine circuit manifold.
10 Digging the trench to lay the brine intake and return pipes.
11 Drilling and building walls for the intake and return pipes.
12 Laying and connecting the intake and return pipes in the trenches from the circuit manifold to the heat pump.
13 Filling, rinsing through and removing air from the borehole heat exchangers with brine, a glycol-water mixture.
14 Isolating the brine pipes within the building to avoid the formation of condense water.
15 Applying for permission from the water authorities and applying for any permissions required to direct any excess water from the drilling into a canal.
16 Removing material extracted from the borehole.

5.4.3 Planning example for a brine-water heat pump system with borehole heat exchangers

The heat output for a detached house was calculated in Section 5.2.5, 'Sample calculation of the heat output for a detached house in accordance with the German Renewable Energies Heat Act':

$$P_{H'} = P_H + P_{HW}$$

$$= 10.4 \text{ kW} + 1.4$$

$$= 11.8 \text{ kW}$$

Assuming a blocking period of four hours:

$$P_{H''} = P_{H'} \times S$$

$$= 10.4 \text{ kW} \times 1.2$$

$$= 12.48 \text{ kW}$$

The next step is to select the right heat pump. The manufacturer offers two brine-water heat pumps with the following output. You choose:

1   a heat pump with a heat output of 12.4 kW; or
2   a heat pump with a heat output of 15.0 kW.

Table 5.12 From a GeoMax® heat pump brochure

Brine-water heat pump, standard reference value: B0/W35

Model		Heat output	Electrical output	COP
GeoMax® SW	4,600	4.6 kW	1.1 kW	4.1
GeoMax® SW	5,900	5.9 kW	1.4 kW	4.1
GeoMax® SW	7,000	7.0 kW	1.7 kW	4.2
GeoMax® SW	8,300	8.3 kW	2.0 kW	4.2
GeoMax® SW	10,300	10.3 kW	2.5 kW	4.2
GeoMax® SW	12,400	12.4 kW	3.0 kW	4.2
GeoMax® SW	15,000	15.0 kW	3.4 kW	4.4
GeoMax® SW	16,800	16.8 kW	3.8 kW	4.4

Width: 700 mm; height: 900 mm; depth: 450 mm

If the customer chooses not to take advantage of the special tariff offered by their utility (something we do not recommend) or if the utility offers no such tariff, then the smaller heat pump with a heat output of 12.4 kW is sufficient, because the calculated heat output excluding blocking periods is:

$$P_{H'} = 11.8 \text{ kW}$$

Where the utility blocking periods need to be taken into account, then the larger heat pump with a heat output of 15 kW is recommended. As this is a 'borderline' case, it may be worth considering if the heat pump could work with a heat output of 12.4 kW.

Here, the abstraction capacity needs to be calculated:

$$P_C = P_{H'} \times 0.8$$
$$= 15.0 \text{ kW} \times 0.8$$
$$= 12 \text{ kW}$$

Now the dimensions of the borehole heat exchangers must be calculated.

It is important that borehole heat exchangers are sufficiently large. The length of the borehole heat exchangers are calculated thus:

$$L = P_C / P_{BHEspec}$$

The value for the specific abstraction capacity per drilled metre $P_{BHEspec}$ can be taken from the table for the specific abstraction capacity per drilled metre. This assumes prior knowledge of the geological ground conditions. Generally, the ground consists of several layers, and the specific abstraction capacity of each layer must be taken into account. This information is available from the local geological service. Many well engineers also have access to this information.

In our example, the ground has the following characteristics:

- Upper layer to 20 m is water-bearing sand and medium-grained gravel: specific abstraction capacity 55–65 W/m. We will take the average value: 60 W/m.
- Lower layer, from 20 m depth, is solid rock with a specific abstraction capacity of 35–45 W/m. To be on the safe side, we take the lower value: 35 W/m.

Now the required length of the borehole heat exchangers can be calculated as follows:

$$L_1 = 20 \text{ m}$$

In the upper layer, the cooling capacity per heat exchanger will be:

$$P_{C1} = L_1 \times P_{BHE1spec}$$
$$= 20 \text{ m} \times 60 \text{ W/m}$$
$$= 1{,}200 \text{ W}$$

If we assume n = 4 borehole heat exchangers, then a total of 4,800 W will be drawn from the upper layer.

Consequently, the lower layers must produce an output of:

$$P_{C2} = P_C - P_{C1} \qquad\qquad L_2 = (P_{C2} / P_{BHE2spec}) / n$$
$$= 15{,}000 \text{ W} - 4{,}800 \text{ W} \qquad\qquad = (10{,}200 \text{ W} / 35 \text{ W/m}) / 4$$
$$= 10{,}200 \text{ W} \qquad\qquad = 72.9 \text{ m}$$
$$\rightarrow \quad L = L_1 + L_2$$
$$= 20 \text{ m} + 73 \text{ m}$$
$$= 93 \text{ m}$$

In this example, it makes sense to drill three borehole heat exchangers with a length of 93 m or 95 m. However, the geological characteristics may make it necessary to drill several borehole heat exchangers of a shorter length. This should be discussed with the well engineer and will depend upon the particular geological conditions.

TIP

- Usually, the length of the borehole heat exchangers will be determined by the well engineer on the basis of the calculated cooling capacity.

- The number of borehole heat exchangers must be agreed with the well engineer before selecting the brine circuit manifold.

Figure 5.24 Illustrations of drilling equipment

The brine circuit manifold shown below is designed for two double U-pipe borehole heat exchangers, or four simple U-pipe borehole heat exchangers.

Figure 5.25 Brine circuit manifold

The customer would also like to use his heat pump system for cooling. Now comes the choice of 'free cooling', or using a reversible heat pump. Reversible heat pumps require power to drive the compressor, something not required with free cooling. This makes free cooling the better alternative, both economically and ecologically. Therefore, here we choose the more environmentally friendly solution – free cooling.

The following diagram shows a heat pump system with borehole heat exchangers and free cooling:

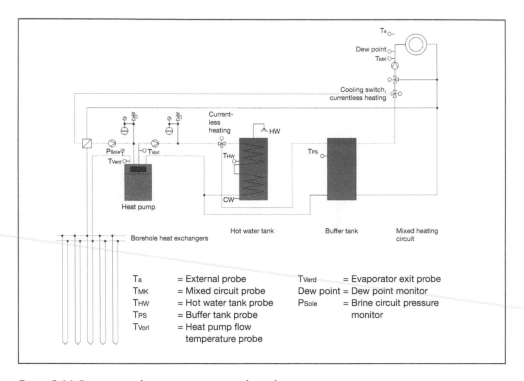

T_a	= External probe		T_{Verd}	= Evaporator exit probe
T_{MK}	= Mixed circuit probe		Dew point	= Dew point monitor
T_{HW}	= Hot water tank probe		P_{Sole}	= Brine circuit pressure
T_{PS}	= Buffer tank probe			monitor
T_{Vorl}	= Heat pump flow temperature probe			

Figure 5.26 Brine-water heat pump system with cooling

TIP

It is recommended that the geological characteristics are investigated prior to building larger developments and so-called heat pump estates. The Geothermal Response Test (GRT) can provide important information for long-term projections of thermal conductivity and abstraction capacity.

Thermal borehole resistance is the thermal conductivity of the grouting material between exchanger and borehole wall. Only a homogenous grout with high level of conductivity will allow the optimal transfer of heat to the heat exchanger.

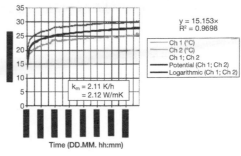

Figure 5.27
Geothermal Response Test

Figure 5.28
Temperature curve of a Geothermal Response
Test

5.4.4 Planning ground heat exchangers

Ground heat exchangers must also be sufficiently large and installed at the correct depth:
min. 1.2 m, max. 2 m.

The following diagram shows that the ground cools most strongly just when the greatest
amount of thermal energy is needed and drawn from the ground. Therefore, it is extremely
important that the ground heat exchangers are carefully planned and sufficiently generously
dimensioned.

The ground heat exchanger consists of horizontally laid PE pipes of DN 20 to DN 32
mm (usually DN 20), according to the ground characteristics and heat pump output, and
with the pipes separated by a distance of 0.3 to 0.8 m. The loops must all be the same
length (often 100 m – the standard delivery length). The longer each loop, the greater the
pressure loss and higher the circulation pump output. The loops must be of equal length
and connected to the circuit manifold according to Tichelmann, to ensure that the brine
follows evenly through all the loops. When installing the pipes, care must be taken that
they are not damaged or impacted by stones or other sharp artefacts. If necessary, the pipes
can be laid down in a bed of fine sand or soft earth.

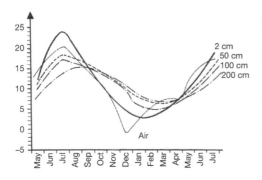

Figure 5.29
Temperature curves at the various depths

Figure 5.30
Installation of ground heat exchangers

The follow reference values apply for the abstraction capacity of ground heat exchangers P_{Cspec}:

Table 5.13 Specific abstraction capacity for ground heat exchangers

Soil type	Abstraction capacity at 1,800 h/a (W/m²)	Abstraction capacity at 2,400 h/a (W/m²)
dry sand, gravel	9–12 W/m²	6–10 W/m²
wet sand, gravel	15–20 W/m²	12–16 W/m²
dry clay, loam	20–25 W/m²	15–20 W/m²
wet clay, loam	25–30 W/m²	20–24 W/m²
water-bearing soil	30–40 W/m²	25–32 W/m²

TIP

The standard reference values given in German guideline VDI 4640 are less comprehensive. The specific abstraction values can fluctuate greatly depending upon soil conditions and any inhomogeneity.

Now we calculate the area required for the heat exchangers:

$$A_{GHE} = P_C / P_{Cspec}$$

where A_{GHE} = area covered by ground heat exchangers [m²]
P_C = cooling capacity [kW]
P_{Cspec} = area-specific abstraction capacity [W/m]

TIP

Ground heat exchangers must be laid at a distance of over 1 m away from the neighbouring plot of land, and more than 0.7 m away from supply lines.

Figure 5.31 is an aid to determining the maximum distance between the loops d_L:

As stated in VDI 4640.9, the maximum distance between the loops for an operating time of 1,800 h/a can also be calculated with the following equation:

$$d_{L18} = 1.17 \text{ m} - P_{Cspec} \times 0.017 \text{ m}^3/\text{W}$$

And for an operating period of 2,400 h/a, the distance can be calculated as follows:

$$d_{L24} = 1.17 \text{ m} - P_{Cspec} \times 0.021 \text{ m}^3/\text{W}$$

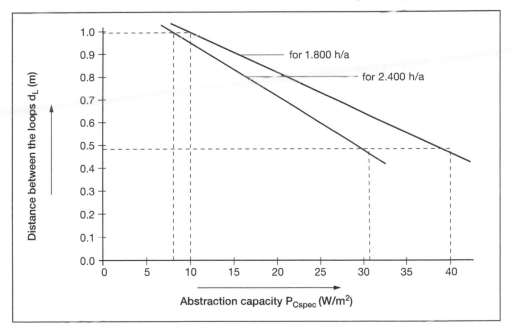

Figure 5.31 Distance between the loops in relation to abstraction capacity

---

TIP

When there is uncertainty, there is no disadvantage to a closer installation. That means that, when in doubt, the area can be somewhat larger and the distance between the loops somewhat smaller.

It is also possible to combine the often more economical ground heat exchangers with significantly more effective borehole heat exchangers. If the installation layout allows, borehole heat exchangers can be subsequently installed at a later date.

5.4.5 Planning example for a brine-water heat pump system with ground heat exchangers

The abstraction capacity of ground heat exchangers is calculated in the same manner as for borehole heat exchangers.

Therefore, for our detached house, the ground heat exchangers again have an abstraction capacity of 12 kW.

Now the ground heat exchangers must be dimensioned. The ground around the planned house consists of a lightly moist mixture of clay and sand. The annual operating time is taken as 1,800 h/a. Taking the table for the specific abstraction capacity as our reference, an average value of 30 W/m² is selected for the specific abstraction capacity. Consequently, the area required for the ground heat exchangers is:

$$A_{GHE} = P_C / P_{Cspec}$$
$$= 12 \text{ kW} / 20 \text{ W/m}^2$$
$$= 12{,}000 \text{ W} / 22.5 \text{ W/m}^2$$
$$= 533 \text{ m}^2$$

where $P_{Cspec} = 22.5 \text{ W/m}^2$

TIP

Now we must check if the plot of land is large enough!

This is often the reason that ground heat exchangers cannot be installed, particularly in urban areas.

Now we must calculate the distance between the loops d_{L18}:

$$d_{L18} = 1.17 \text{ m} - P_{GHEspec} \times 0.017 \text{ m/w}$$
$$= 1.17 \text{ m} - 22.5 \text{ W/m}^2 \times 0.017 \text{ m/w}$$
$$= 0.79 \text{ m}$$

Or by referring to the following diagram:

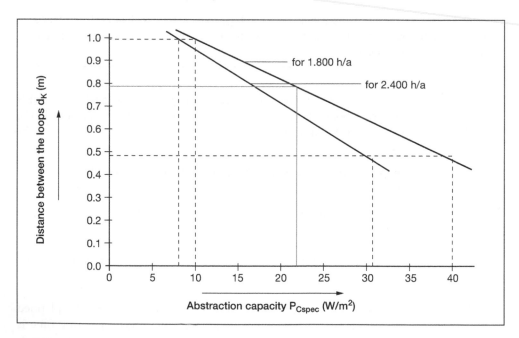

Figure 5.32 Calculating the distance between the loops for an abstraction capacity of 30 W/m²

The diagram shows that as the specific abstraction capacity of the upper soil level increases, the distance between the loops decreases in a linear relation. This is also described by the equation.

In order to calculate the number of loops, the individual loops must be distributed over the plot of land.

TIP

It is helpful to sketch each loop on the layout plan.

Laying the ground heat exchangers:

First, the borders of the plot must be marked, and then the ground heat exchangers sketched in.

Care must be taken not to place the loops too closely together in order to avoid freezing.

When planning the layout of the heat exchanger area, the available area and the delivery lengths of the PE pipes must be considered. Here, we assume a delivery length of 200 m. The 533 m² plot is arranged in an area of roughly 30 m × 18 m. Where the loops are placed 0.8 m apart, at least 12 heat exchanger loops are needed. You are planning four heat exchanger circuits, each with a total length of 150 m. It is recommended that the heat exchanger loop layout is sketched, for example:

1   Laid according to the Tichelmann system.

2   Here, four ground heat exchanger circuits are laid out.

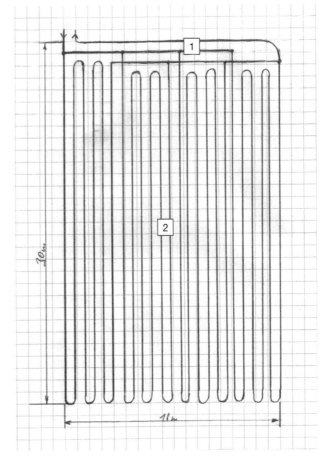

Figure 5.33
Layout of the ground heat exchangers

If the individual loops are not connected to a manifold that enables flow control, then it is essential that the loops are laid out according to the Tichelmann system, as shown here. It is also important that each loop is the same length. Both these measures are necessary to ensure that the flow is the same in all the loops.

It is also a good idea to connect the individual loops to a manifold with flow splitters that enable the flow through each loop to be compared and controlled. They also help localise any undesired disruptions, especially when the heat pump is being started up for the first time and during initial operation.

TIP

When laying the connection lines (e.g. to a brine circuit manifold), it is essential that these are not placed too closely together near the manifold to avoid overcooling (freezing) of the ground in this area.

It is always a good idea to sketch the layout of the heat exchangers and the PE pipe loops on paper first.

TIP

During cold weather or when laid closely together, tightly looped pipes can be difficult to lay. An option is to allow the loop bends to take up more space, or to allow them to overlap.

5.4.6 Energy mats – an alternative to ground heat exchangers?

Energy mats for heat pumps are mats interwoven with closely laid, thin plastic pipes. Due to the dense layout of the pipes, they claim to draw more thermal energy out of the ground per given area, even when installed partly under building structures. This makes no sense in physical terms because, as we have seen, only a set quantity of heat is available in any given area (see Section 5.4.4, 'Planning ground heat exchangers', Table 5.13), the specific amount depending on the particular characteristics of the ground. The method by which the heat is extracted from the ground is irrelevant; the same calculation basis is used for ground heat exchangers and energy mats, even though they have closely packed, thin, parallel plastic pipes. A further consideration is that these thin plastic pipes are more easily damaged.

The following observation shows that energy mats cannot be a true alternative to ground heat exchangers. From underfloor heating technology, we know that, for example, underfloor piping is more densely laid in bathrooms because the floor needs to be warmer than other rooms heated in this manner. By packing the pipes more closely together, the overall surface area of the underfloor piping is greater. Therefore, more heat can be emitted from this larger surface area, even though the flow temperature is the same as for other rooms in the house. Naturally, heat output increases as the surface temperature increases. However, the heat is coming from the boiler and we can assume that the boiler has sufficient capacity to provide this heat.

The energy mats operate similarly, but in reverse. More thermal energy is extracted from the ground because the pipes are laid more closely together. The heat pump abstracts the energy it needs, causing the temperatures at the heat source to fall. Because this heat extraction is occurring in a concentrated area, the ground in this area is cooled strongly and the brine temperature falls. The ground can even freeze and, in extreme cases, results in the heat pump being shut down (low pressure fault).

Furthermore, the fall in the brine temperature reduces the degree of efficiency and thus level of performance for the heat pump system as a whole.

There are also energy mats available for installing on the outside of houses. In some, ambient air flows through the pipes. These energy mats use solar energy as well as the thermal energy in the ambient air. Both these energy sources are subject to significant fluctuations, and this in turn leads to fluctuations in the performance of the heat pump system.

These observations show that so-called energy mats, especially when laid in the ground, are not a real alternative to ground heat exchangers. They are not more effective, and can be more expensive.

5.4.7 Planning geothermal baskets

Geothermal baskets are similar to ground heat exchangers as they are also embedded in the ground, close to the surface.

According to the manufacturers, at least 10 m² of open ground is necessary for each basket, and a minimum distance of 4 m must be maintained between each basket.

The geothermal baskets need to be embedded in ground that is water-retentive, if not water-bearing.

As with ground heat exchangers, the heat is transferred from the ground via the surface of the geothermal basket, as well as over its length of around 2 to 2.5 m, as with borehole heat exchangers. When we add the abstraction capacity, this gives the following:

Table 5.14 Abstraction capacity of geothermal baskets

dry sand, gravel	120–180 W/basket
wet sand, gravel	190–240 W/basket
dry clay, loam	240–310 W/basket
wet clay, loam	310–380 W/basket
water-bearing soil layers	390–460 W/basket

Figure 5.34 Geothermal basket

Because of their proximity to the surface (seasonal variations in temperature), geothermal baskets, like ground heat exchangers, are not suitable for free cooling.

5.4.8 Ditch heat exchangers

As the term suggests, ditch heat exchangers are laid in ditches, usually with horizontal lengths of PE pipes. The water flowing through these ditches can help heat pump systems with ditch heat exchangers achieve high performance levels. Care must be taken that the abstraction capacity is sufficient, and the fluctuating temperatures of the (flowing) water must be considered.

5.4.9 Roof absorbers, energy fences, solid absorbers, etc.

Other potential heat sources include roof absorbers, energy fences and solid absorbers. Energy fences are long stretches of absorbers, installed like fences: posts are set in the ground and the PE pipes tied, horizontally, to the line of posts.

Energy fences can also be built into walls, and here we talk of solid absorbers (e.g. solid concrete blocks in which PE pipes are embedded).

Heat is extracted from the environment simply by cooling the ambient air. This extracted heat is transferred to the heat pump via the brine circuit. This type of heat pump is effectively a combination of a brine-water heat pump and an air heat pump without a ventilator. Every air-water heat pump has a ventilator to ensure sufficient air circulation and thus energy supply. This is not the case here.

Exact planning is problematic. If an absorber is too small, the brine temperature can fall far below 0 °C. This in turn leads to a low level of efficiency and, in extreme cases, to a low pressure fault and the shutting down of the heat pump. These absorbers often freeze during heating periods in which external temperatures are low.

Here, we will look at a quick example: a roof absorber with a surface area of around 20 m² and an area of 130 m² that requires heating. The heat pump has an output of 8 kW. The cooling capacity is around 6.4 kW. This is extracted from the environment via the roof absorber. If this roof absorber is the only heat source, then we can assume that it is too small, and on cold days the heat pump works less efficiently. The opposite is the case when the sun shines on the roof absorber; here, the absorber will become almost so hot that the heat pump no longer needs to operate.

This is the case for energy fences and solid absorbers, etc. If adequately dimensioned and installed so that both sun and wind can provide sufficient heat, then they can be used to drive a heat pump.

As brine temperatures are low on cold days, heat pump performance on cold days is correspondingly low. The opposite is true on warm, sunny days in spring and autumn. In summary: these systems offer the lowest level of performance precisely when the greatest amount of heat is needed for heating; their level of performance is only good when less heat is needed and they have been correctly installed.

Let us look again at our example: brine flows through the roof absorber and the external temperature is 0 °C. Because of the relatively small surface area, we can assume that the roof absorber will freeze and the temperature of the brine fed to the heat pump is around −10 °C or even less. Naturally, this has a significant negative impact on the COP. With a flow temperature of 30 °C, the energy efficiency ratio can be calculated as follows:

$$\varepsilon = 0.5 \times T_H / (T_H - T_A)$$

$$= 0.5 \times 303 \text{ K} / (303 \text{ K} - 263 \text{ K})$$

$$= 3.79$$

A brine-water heat pump working with the same flow temperature and a brine temperature according to DIN EN 14511-2 would have an energy efficiency ratio of:

$$\varepsilon = 0.5 \times T_H / (T_H - T_A)$$

$$= 0.5 \times 303 \text{ K} / (303 \text{ K} - 273 \text{ K})$$

$$= 5.05$$

This example shows very clearly that it is least efficient when the heat pump output needs to be at its highest.

Some suppliers of such absorbers recommend an additional ground heat exchanger, which can then be somewhat smaller: the ground heat exchanger used together with the absorber ensures more even heat source temperatures. This combination is advantageous when there is too little space to install ground heat exchangers alone.

So-called 'energy fences' should be as unobstructed as possible to improve the transfer of heat from circulating air on windy days. In the end, this type of heat pump technology is similar to an air heat pump without a ventilator, and so the natural airflow needs to be as unimpeded as possible. Energy fences covered by plants are neither good for the energy fence, nor for the plants.

The only benefits offered by this absorber technology are that it can be used as an additional method of heat extraction in an area too small for ground heat exchangers alone, and that it causes no noise emissions, as is the case with air evaporators.

5.4.10 Lake absorber/river absorber

Due to its very high specific heat capacity of 4.182 kWs/kgK, water is an excellent source of heat. Therefore, lakes, and especially flowing bodies of water, are outstanding heat sources. This is clearly demonstrated if we look at the Niers River: the Niers is fed by numerous sewage purification plants, making its temperature higher than normal. Cooling would therefore be desirable. We have the following data for the river:

Volume flow of the Niers: $V = 5 \text{ m}^3/\text{s}$ to $10 \text{ m}^3/\text{s}$
Min. measured temperature: $T_{min} = 3 \text{ °C}$

As the Niers never freezes, we can assume the most unfavourable values will be:

Min. volume flow of the Niers: $V_{min} = 5 \text{ m}^3/\text{s}$
Max. cooling: $\Delta T_{max} = 3 \text{ °C} = 3 \text{ K}$

For calculating heat quantity:

$$Q = m \times c \times \Delta T$$

where Q = heat quantity [KWs]
m = mass [kg]
c = 4.182 kJ/kgK = 4.182 kWs/kgK = 4.182 kWs / tK = specific heat capacity
ΔT = temperature difference [K]

For calculating performance:

$$P = V \times c \times \Delta T$$

where   P = performance [kW]
        V = volume flow [m³/s]

This gives us the potential abstraction capacity, or cooling capacity PC, under the least favourable conditions:

$$P_C = 5 \text{ t/s} \times$$

$$= 4.182 \text{ kWs / tK} \times 3 \text{ K}$$

$$= 62,730 \text{ kW}$$

This corresponds to a heat output of around 78,400 kW!

The Niers, a small river on the Lower Rhine, offers sufficient energy to operate numerous heat pumps along the river. There is sufficient energy to heat around 9,800 houses built to the current standards for new developments, each with around 150 m² of living space and designed for four occupants, in a completely environmentally friendly and very inexpensive manner. Using heat drawn from the environment, many more than 10,000 houses could certainly be supplied with this absolutely environmentally friendly form of energy. Germany has very many small and larger rivers that would make excellent heat sources. They are just waiting to be used!

However, it is not so simple to plan heat pumps where the heat source is a lake. The first step is to differentiate between standing and flowing bodies of water, and precise hydrological and geological calculations are required.

It is a real shame that so little use is made of such energy sources, and this is often because public bodies find so many obstacles to issuing permission for them to be used in this way. In the case of the River Niers, connecting heat pumps would be doubly beneficial because the Niers is too warm; heat pumps would both reduce CO_2 emissions and cool the river.

Now we need to choose between two options: to use the water directly, by allowing it to flow into the heat pump's evaporator, or to extract the heat via an intermediate circuit. If we wish to use the water directly, then we need to consider the water quality (suspended matter, corrosiveness). A good filtration system is vital. For indirect use, PE pipes are laid in the lake or river. It is important that they cannot get damaged (e.g. by changes to the river bed caused by flowing water).

We must also consider environmental protection, river use, general usage, dangers, etc.

With regard to environmental protection, when correctly dimensioned the heat pump will help reduce CO_2 emissions. However, an accident (leak) could result in a water-glycol mixture entering the River Niers. In this case, the problem is relativised as the Niers is a waste water river, fed by several sewage purification plants.

The Niers is also used as a recreational river for paddling. Therefore, the pipes would need to be laid on the riverbed in a manner that offers them protection, and be strong enough to almost exclude the possibility of damage from paddles, etc. The only danger is sabotage. In larger rivers, the impact of anchors must be considered, although this can be avoided by marking sections of riverbank where anchors may not be used.

Another benefit of this energy source is that the average temperature is certainly above 0 °C. This further increases the efficiency of the heat pump system as measured by the seasonal performance factor (SPF). That means that these heat pump systems also work more efficiently.

These are opportunities that should be used as heat pumps become increasingly popular.

5.5 Planning an air-water heat pump system

At first glance, planning an air-water heat pump system seems very easy. You calculate the required heat output, browse through the manufacturer's brochures and select the right heat pump. There is no need to plan the connection to a heat source (wells, borehole or ground heat exchangers). Be careful – it is not that easy!

Figure 5.35 Air-water heat pump

Experience shows that there is an almost linear relationship between falling heat pump output, falling external air temperatures and increasing power consumption.

Figure 5.36 shows the heat output of four different air-water heat pumps, depending on external temperature.

The diagram shows that output falls as external temperatures fall. Thus the output is lowest at times when the greatest heat output is required. The diagram also shows that the output of an air-water heat pump increases as external temperatures rise.

So we have two options:

1 to dimension the heat pump based on the coldest regional temperature (e.g. −12 °C); or
2 to choose bivalent operation and set a bivalence point.

Both options must be based on the lowest expected standard external temperature as set out in DIN EN 12831. Depending on exact location, in Germany this varies between −10 °C and −20 °C. This level of fluctuation is considerable and should not be ignored.

IMPORTANT

A buffer tank is recommended for air-water heat pumps, because as external temperatures increase, air-water heat pumps increasingly tend to turn on and off at frequent intervals.

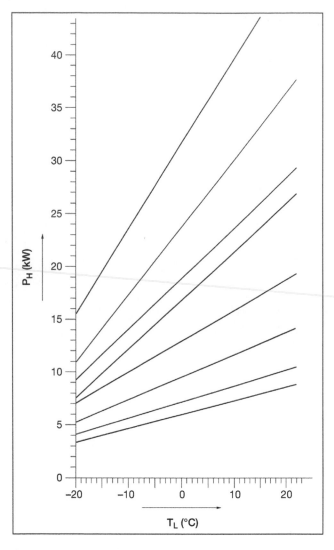

Figure 5.36 COP for various air-water heat pumps depending upon external temperature

5.5.1 Dimensioning an air-water heat pump system for the lowest temperature

If the air-water heat pump is to be dimensioned according to the lowest temperature, then the local standard external temperature must be determined in accordance with DIN EN 12831. The air-water heat pump output can then be calculated (or the manufacturer consulted) using this standard external temperature as the basis.

5.5.2 Calculating the bivalence point of an air-water heat pump system

If an air-water heat pump is dimensioned according to the standard external temperature, then it will be very large and over-dimensioned for the transitional periods. This is why air-water heat pumps are often operated in bivalent mode. This requires that a bivalence point is first established (i.e. a temperature is set below which auxiliary heating, usually an electrical heating element, cuts in). In turn, this reduces the SPF, and therefore the COP, of the heat pump.

The following examples show clearly how the bivalence point is set.

5.5.3 Example of planning an air-water heat pump system

In Section 5.2.5, 'Sample calculation of the heat output for a detached house in accordance with the German Renewable Energies Heat Act', the heat output for a detached house was calculated assuming a utility blocking time of four hours:

$$P_{H''} = P_{H'} \times S$$
$$= 10.4 \text{ kW} \times 1.2$$
$$= 12.48 \text{ kW}$$
$$= 12.5 \text{ kW}$$

Now we need to select the most suitable heat pump. As we are dealing with a new and densely populated housing development, the heat pump needs to be installed indoors. The standard external temperature is −12 °C.

First, we select the following models from the manufacturers' technical data sheets:

Table 5.15 Excerpt from the data sheets

Technical data			
Type	WPL 13 E basic unit	WPL 18 E basic unit	WPL 23 E basic unit
Weight	210 kg	220 kg	225 kg
Volume flow on the heating side	1.0 m³/h	1.2 m³/h	1.4 m³/h
Volume flow on the heat source side	3,200 m³/h	3,500 m³/h	3,500 m³/h
Connection on heating side	G 1¼″	G 1¼″	G 1¼″
Starting current	30 A	30 A	30 A
Hot water temperature differential	4.5 K	4.5 K	4.5 K
Heat output at A2/W35	8.1 kW	11.3 kW	14.8 kW
COP at A2/W35	3.4	3.7	3.5
Heat output at A-7/W35	6.6 kW	9.6 kW	13 kW
COP at A-7/W35	3.0	3.2	3.1

Output data according to DIN EN 14511

Now we must determine the bivalence point, either by means of a graph, or mathematically. First, we enter the line for the heating characteristics of the building, depending upon external temperature:

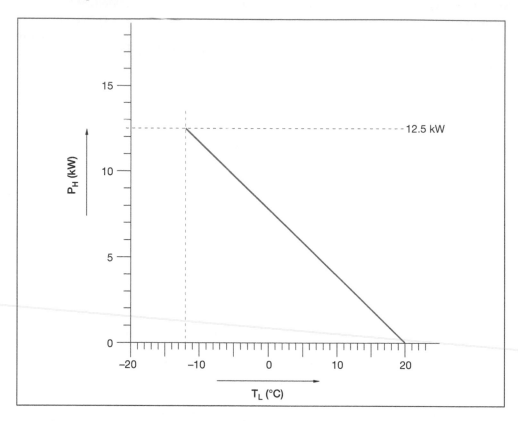

Figure 5.37 Heat output of the building

To calculate the bivalence point graphically, we first need to enter the two operating points (i.e. the heat pump output at 2 °C: 14.8 kW, and the output at –7 °C: 13 kW) (black vertical dotted lines in Figure 5.38). When we draw a line through the intersection points, we get the characteristic describing the output of the heat pump according to external temperature.

The next step is to determine the characteristic that describes the building's average required heat demand. Here again, we have two intersecting points: 12.5 kW is required at –12 °C, and at 20 °C no more heat pump output is required (red dotted lines). If we connect the points, we get a line that describes the output needed to cover heat demand. Where these two lines intersect, we have the bivalence point. (See Figure 5.38.)

The bivalence point can also be determined mathematically using linear equations f(x). The following equation is used to describe the line:

$$f(x) = a_0 + a_1 x$$

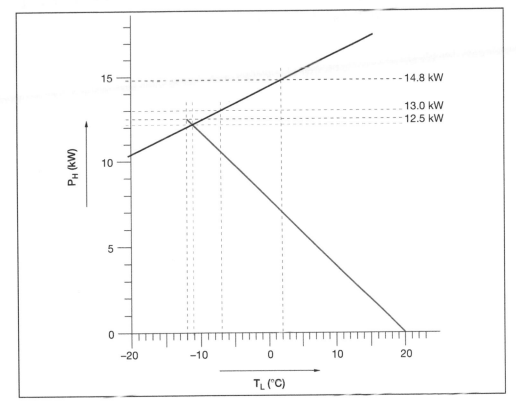

Figure 5.38 Determining the bivalence point

where f(x) = function of x
a₀ = zero at f(x = 0)
a₁ = increase

The following applies:

$$f(x) = P_{HP}(T) = a_0 + a_1 x$$

where $P_{HP}(T)$ = heat pump output depending on external temperature T = x

At 2 °C, the heat pump has an output of 14.8 kW. Thus:

$$P_{HP}(T = 2\ °C) \quad = a_0 + a_1\ 2\ °C = 14.8\ kW \tag{1}$$

At –7 °C, the heat pump has an output of 13.0 kW. Thus:

$$P_{HP}(T = -7\ °C) \quad = a_0 + a_1\ (-7\ °C) = 13.0\ kW \tag{2}$$

$$\rightarrow \quad a_0 + a_1\ (-7\ °C) \quad = 13.0\ kW$$

$$\rightarrow \quad a_0 = 13.0\ kW - (a_1\ (-7\ °C)) = 13.0\ kW + a_1\ 7\ °C \tag{3}$$

Where this equation (3) is entered into the first (1), then:

$$P_{HP}(T = 2\ °C) = a_0 + a_1\ 2\ °C = 14.8\ kW$$
$$= 13.0\ kW + a_1\ 7\ °C + a_1\ 2\ °C$$
$$= 14.8\ kW$$
$$= 130\ kW + a_1\ 9\ °C$$
$$= 14.8\ kW$$

$$\rightarrow \quad a_1\ 9\ °C = 14.8\ kW - 13.0\ kW = 1.8\ kW$$

$$\rightarrow \quad a_1 = 0.2\ kW/°C \tag{4}$$

Where this equation (4) is entered into (3), then:

$$a_0 = 13.0\ kW + a_1\ 7\ °C$$
$$= 13.0\ kW + 0.2\ kW/°C \times 7\ °C$$
$$= 13.0\ kW + 1.4\ kW$$
$$a_0 = 14.4\ kW \tag{5}$$

Now for the equations (4) and (5):

$$f(x = T) = P_{HP}(T) = a_0 + a_1\ T$$

$$P_{HP}(T) = 14.4\ kW + 0.2\ kW/°C \times T$$

where   T = external temperature [°C]

In the same way, the equation for the required heat output, depending on the external temperature, can be calculated.

Here again:

$$f(x) = a_0 + a_1 x$$

At –12 °C, the maximum heat output of P_H = 12.48 kW is needed, and at 20 °C heating is no longer required (i.e. P_H = 0 kW). Thus:

$$f(x = T) = P_H(T)$$
$$= a_0 + a_1\ T$$

$$f(x = T = -12\ °C) = P_H(T = -12\ °C)$$
$$= a_0 + a_1\ (-12\ °C)$$
$$= 12.48\ kW \tag{6}$$

and $f(x = T = 20\ °C) = P_H(T = 20\ °C)$

$$= a_0 + a_1\ 20\ °C$$

$$= 0\ kW \tag{7}$$

$\rightarrow$ $a_0 = -a_1\ 20\ °C \tag{8}$

Where this equation (8) is entered back into (6), then:

$f(x = T = -12\ °C) = P_H(T = -12\ °C)$

$= a_0 + a_1\ (-12\ °C)$	$= 12.48\ kW$
$= -a_1\ 20\ °C + a_1\ (-12\ °C)$	$= 12.48\ kW$
$= -a_1\ 20\ °C - a_1\ 12\ °C$	$= 12.48\ kW$
$= -a_1\ (20\ °C + 12\ °C)$	$= 12.48\ kW$
$= -a_1\ 32\ °C$	$= 12.48\ kW$

$\rightarrow$ $a_1 = -12.48\ kW\ /\ 32\ °C = -0.39\ kW/°C \tag{9}$

$\rightarrow$ $a_0 = a_1\ 20\ °C$

$$= - (-0.39\ kW/°C) \times 20\ °C$$

$$= 0.39\ kW/°C \times 20\ °C$$

$$= 7.8\ kW \tag{10}$$

Therefore, for the required heat output P_H:

$$P_H(T) = 7.8\ kW - 0.39\ kW/°C \times T \tag{11}$$

where   T = external temperature [°C]

A control calculation shows that the equation is correct:

1   Heat pump output at 2 °C:

$$P_{HP}(T = 2\ °C) = 14.4\ kW + 0.2\ kW/°C \times 2\ °C = 14.8\ kW$$

Heat pump output at –7 °C:

$$P_{HP}(T = -7°C) = 14.4\ kW + 0.2\ kW/°C \times (-7\ °C) = 13.0\ kW$$

And:

2   Heat output needed by the building at –12 °C:

$$P_H(T = -12\ °C) = 7.8\ kW - 0.39\ kW/°C \times (-12\ °C)$$

$$= 7.8\ kW + 4.68\ kW$$

$$= 12.48\ kW$$

Heat output needed by the building at 20 °C:

$$P_H(T = 20\ °C) = 7.8\ kW - 0.39\ kW/°C \times (20\ °C)$$

$$= 7.8\ kW - 7.8\ kW$$

$$= 0\ kW$$

The bivalence point can be determined using these equations. The lines cross at the bivalence point – here, they have the same output (P_{Biv}) and the same temperature (T_{Biv}). Thus:

$$P_{HP}(T_{Biv}) = 14.4\ kW + 0.2\ kW/°C \times T_{Biv}$$

$$= P_H(T_{Biv})$$

$$= 7.8\ kW - 0.39\ kW/°C \times T_{Biv} \qquad (12)$$

→ $14.4\ kW + 0.2\ kW/°C \times T_{Biv} = 7.8\ kW - 0.39\ kW/°C \times T_{Biv}$

→ $6.6\ kW + 0.2\ kW/°C \times T_{Biv} = -0.39\ kW/°C \times T_{Biv}$

→ $6.6\ kW = -0.39\ kW/°C \times T_{Biv} - 0.2\ kW/°C \times T_{Biv}$

→ $6.6\ kW = (-0.39\ kW/°C - 0.2\ kW/°C) \times T_{Biv}$

→ $6.6\ kW = -0.959\ kW/°C \times T_{Biv}$

→ $T_{Biv} = 6.6\ kW / (-0.959\ kW/°C)$

→ $T_{Biv} = -11.19\ °C \qquad (13)$

Where the temperature falls below this average external temperature, then the controller should switch on the auxiliary heating.

IMPORTANT

The bivalence point must never diverge too far from the standard external temperature!

Where it diverges too greatly, the auxiliary heating will be turned on too often, and this should be avoided.

In this planning example, the bivalence point lies very close to the standard external temperature of –12 °C, which we determined earlier. If we refer back to the technical data, we could consider using the slightly smaller heat pump instead. Therefore, we need to calculate the bivalence point of the smaller heat pump.

Mathematical calculation of the bivalence point. Again, the following equation applies for describing a straight line for the heat pump output, depending on external temperature:

$$f(x) = a_0 + a_1 x$$

where $f(x)$ = function of x
a_0 = zero
a_1 = increase

The following applies:

$$f(x) = P_{HP}(T) = a_0 + a_1 x$$

where $P_{HP}(T)$ = heat pump output depending on external temperature $T = x$

At 2 °C, the heat pump has an output of 11.3 kW. Thus:

$$P_{HP}(T = 2 °C) = a_0 + a_1 \ 2 °C = 11.3 \ kW \tag{1}$$

At –7 °C, the heat pump has an output of 9.6 kW. Thus:

$$P_{HP}(T = -7 °C) = a_0 + a_1 \ (-7 °C) = 9.6 \ kW \tag{2}$$

$\rightarrow \quad a_0 + a_1 \ (-7 °C) = 9.6 \ kW$

$\rightarrow \quad a_0 = 9.6 \ kW - (a_1 \ (-7 °C)) = 9.6 \ kW + a_1 \ 7°C \tag{3}$

Where this equation (3) is entered into the first (1), then:

$$P_{HP}(T = 2 °C) = a_0 + a_1 \ 2 °C = 11.3 \ kW$$

$$= 9.6 \ kW - a_1 \ (-7 °C) + a_1 \ 2 °C = 11.3 \ kW$$

$$= 9.6 \ kW + a_1 \ 7 °C + a_1 \ 2 °C = 11.3 \ kW$$

$\rightarrow \quad a_1 \ 7 °C + a_1 \ 2 °C = 11.3 \ kW - 9.6 \ kW$

$\rightarrow \quad a_1 \ 7 °C + a_1 \ 2 °C = 1.7 \ kW$

$\rightarrow \quad a_1 \ (7 °C + 2 °C) = 1.7 \ kW$

$\rightarrow \quad a_1 \ 9 °C = 1.7 \ kW$

$\rightarrow \quad a_1 = 0.19 \ kW/°C \tag{4}$

Where we enter equation (4) into equation (3), then:

$$a_0 = 9.6 \ kW + a_1 \ 7 °C$$

$$= 9.6 \ kW + 0.19 \ kW/°C \times 7 °C$$

$$= 9.6 \ kW + 1.3 \ kW$$

$$a_0 = 10.9 \ kW \tag{5}$$

Now with the equations (4) and (5):

$$f(x = T) = P_{HP}(T) = a_0 + a_1 T$$

$$P_{HP}(T) = 10.9 \ kW + 0.19 \ kW/°C \times T \tag{6}$$

where T = external temperature [°C]

The equation for required heat output dependent on the external temperature is again:

$$P_H(T) = 7.8 \text{ kW} - 0.39 \text{ kW/°C} \times T \tag{7}$$

where T = external temperature [°C]

The equations above allow us to determine the bivalence point (where the two lines cross) and therefore the temperature (T_{Biv}) at the bivalence point. Thus:

$$P_{HP}(T_{Biv}) = 10.9 \text{ kW} + 0.19 \text{ kW/°C} \times T_{Biv}$$
$$= P_H(T_{Biv}) = 8 \text{ kW} - 0.94 \text{ kW/°C} \times T_{Biv} \tag{8}$$

$\rightarrow$ $10.9 \text{ kW} + 0.19 \text{ kW/°C} \times T_{Biv} = 7.8 \text{ kW} - 0.39 \text{ kW/°C} \times T_{Biv}$

$\rightarrow$ $3.1 \text{ kW} + 0.19 \text{ kW/°C} \times T_{Biv} = -0.39 \text{ kW/°C} \times T_{Biv}$

$\rightarrow$ $3.1 \text{ kW} = -0.39 \text{ kW/°C} \times T_{Biv} - 0.19 \text{ kW/°C} \times T_{Biv}$

$\rightarrow$ $3.1 \text{ kW} = (-0.39 \text{ kW/°C} - 0.19 \text{ kW/°C}) \times T_{Biv}$

$\rightarrow$ $3.1 \text{ kW} = -0.958 \text{ kW/°C} \times T_{Biv}$

$\rightarrow$ $T_{Biv} = 3.1 \text{ kW} / (-0.958 \text{ kW/°C})$

$\rightarrow$ $T_{Biv} = -5.3 \text{ °C} \tag{9}$

Where the temperature falls below this average external temperature, the controller should turn on the auxiliary heating.

Determining the bivalence point for the smaller heat pump by means of a diagram is shown in Figure 5.39.

This gives us a bivalence point of $T_{Biv} = -5.3$ °C.

PLEASE NOTE

Here, it is clear that the bivalence point is significantly higher for the smaller heat pump. Consequently, the auxiliary heat is switched on correspondingly earlier.

The bivalence point for the first heat pump lies very close to the standard external temperature. Therefore, if using the larger heat pump, it would be worth considering doing without the auxiliary heating. For the next heat pump in size down, auxiliary heating is certainly required. An even smaller heat pump is certainly not recommended!

For a water-water heat pump or a brine-water heat pump, heat output is largely independent of the external temperature. With a brine-water heat pump, only the lightly fluctuating brine temperatures (the result of the ground cooling at the beginning of the heating period and regenerating in the summer) are perceptible.

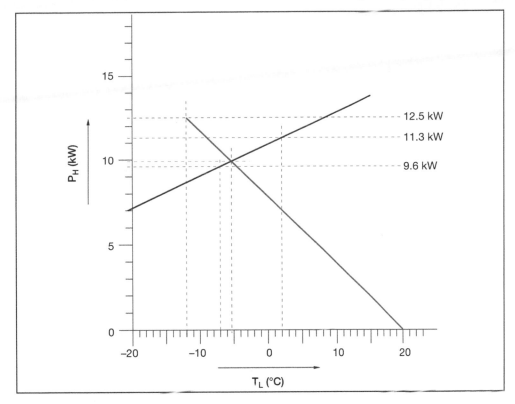

Figure 5.39 Determining the bivalence point

Thus the output of a water-water or brine-water heat pump can almost be assumed to be constant. If one of these heat pumps has a heat output of 10.3 kW, for example, then it is easy to determine the bivalence point. In this example, it is around –2 °C.

Figure 5.42 shows the share of the annual heating output covered by the heat pump for a standard year, depending upon its size, mode of operation, the relationship between the heat pump's heat output and the maximum standard heating load of the building, and the operating method.

TIP

As for standard external temperatures, the standard year is an average value. Do not forget that significant deviations both above and below this standard are possible.

Bivalent alternative operation means that where the temperature falls below the bivalence point, the heat pump shuts down and the auxiliary heating alone generates all the heat output. The auxiliary heating may be driven by oil, gas or pellets, instead of electricity. This method of heating is often used where a heating system is already present and the heat pump is installed at a later date. However, in most cases, this makes absolutely no sense, because when the first heating system becomes defective or reaches its end of life,

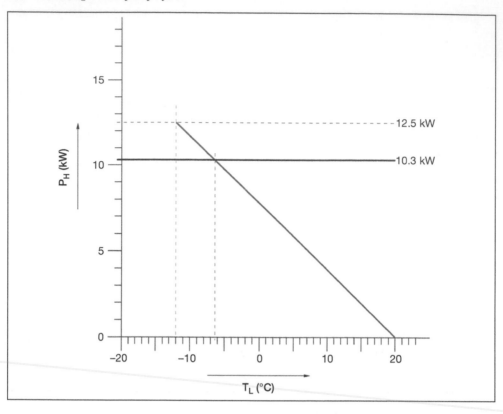

Figure 5.40 Determining the bivalence point where heat output is constant

it must be replaced. Consequently, two heating systems are run in parallel with all the additional maintenance and servicing this entails. This generally makes no economic sense. Bivalent, partly parallel operation means that for temperatures above 0 °C, the heat pump alone takes on the task of heating. Below this temperature, the heat pump operates in parallel with the auxiliary heating system. When the temperature falls further, here −10 °C, the heat pump shuts down completely, for technical reasons. This form of heating is not recommended for small heat pump systems.

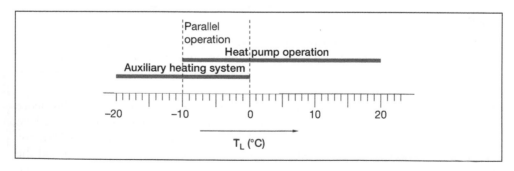

Figure 5.41 Partly parallel operation

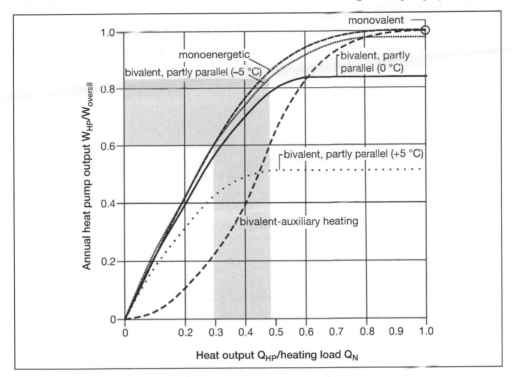

Figure 5.42 Share of annual heat output contributed by the heat pump in accordance with dimensioning and mode of operation

Figure 5.42 shows the share of heating work that the heat pump performs in a standard year W_{HP} in relation to the entire heating load for the year $W_{overall}$ as a function of the heat output of the heat pump Q_{HP} in relation to the maximum heating load Q_N. This diagram allows us to estimate the operating efficiency of the heat pump system when operating in bivalent mode for various forms of operation.

The relationship $W_{HP}/W_{overall}$ multiplied by 100 per cent gives us the heat pump's contribution to the annual heating load in the form of a percentage.

When selecting the bivalence point, care must be taken that it lies above or to the right of the area marked in orange. It certainly should not fall below or to the left of this area; here, we can assume that the heat pump is certainly no longer operating efficiently.

Now we examine the following modes of operation:

Monovalent operation:

When operated in monovalent mode, the heat pump alone supplies the necessary heating load up to standard external temperatures, with no auxiliary heating.

Here:

$$W_{HP} / W_{overall} = Q_{HP} / Q_N = 1$$

Monoenergetic operation:
Where heat pump operation is monoenergetic, the auxiliary electric heating cuts in where the bivalence point is undercut. The heat pump operates at full power, the auxiliary heating only compensating for the heat pump deficit. However, monoenergetic operation can also be bivalent-auxiliary operation where the electric heating takes over the entire work of heating when the bivalence point is undercut.

Most air-water heat pumps work in this way.

Bivalent, partly parallel operation:
Here, the heat pump works on its own until the temperature falls to a set level. Where the temperature falls below this point, the auxiliary heating cuts in and works in parallel with the heat pump. The heat pump continues to work at full power. Only if the temperature falls further, below an even lower minimum level, does the heat pump shut down and the auxiliary heating takes over the entire heating operation.

IMPORTANT

In principle, a water-water or a brine-water heat pump should be operated in mono-valent mode. It makes no sense to operate these heat pumps in bivalent mode.

Let us look again at the two air-water heat pumps. At a standard external temperature of –12 °C, the required standard heating load for the building is 12.5 kW. At a standard external temperature of –12 °C, the larger air-water heat pump has an output of:

$$P_{HP}(T_N) = 14.4 \text{ kW} + 0.2 \text{ kW/°C} \times T$$

$$= 14.4 \text{ kW} + 0.2 \text{ kW/°C} \times (-12 \text{ kW})$$

$$= 14.4 \text{ kW} - 2.4 \text{ kW}$$

$$= 12 \text{ kW}$$

The relationship Q_{HP} / Q_N is calculated as:

$$Q_{HP} / Q_N = 12 \text{ kW} / 12.5 \text{ kW}$$

$$= 0.96$$

Figure 5.43 shows the choice of bivalence point and determining the heat pump's share of the annual heat output.

In order to determine the relationship Q_{HP} / Q_N, we need to calculate and add the heat output of the heat pump at the standard external temperature for Q_{HP}.

This shows that the larger air-water heat pump almost exclusively operates in monovalent mode, with the electrical heating element rarely being switched on (see green dashed lines).

For the smaller heat pump, the auxiliary heating switches on more frequently, and so here we must consider the auxiliary heating's mode of operation. This can incur considerable additional costs (e.g. where the auxiliary operation is in bivalent mode).

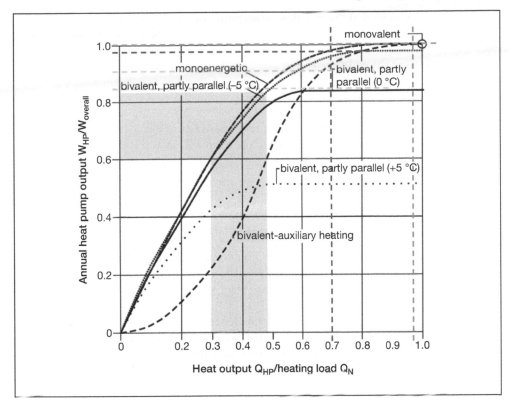

Figure 5.43 Share of annual heat output contributed by the heat pump in accordance with dimensioning and mode of operation

For the heat pump model WPL 18 E, in the best case the monoenergetic operation is:

$$P_{HP}(T) = 10.9 \text{ kW} + 0.19 \text{ kW/°C} \times T$$

$$= 10.9 \text{ kW} + 0.19 \text{ kW/°C} \times (-12 \text{ °C})$$

$$= 10.9 \text{ kW} - 2.28 \text{ kW}$$

$$= 8.62 \text{ kW}$$

The relationship Q_{HP} / Q_N is calculated as:

$$Q_{HP} / Q_N = 8.62 \text{ kW} / 12.5 \text{ kW}$$

$$= 0.969$$

This shows that the smaller air-water heat pump is already operating in bivalent mode, but the electric heating element is switched on more rarely for the smaller heat pump. This heat pump will take over around 97 per cent of the heating load for the building (i.e. on very cold days, the heat pump is working in borderline conditions), and the remaining 3 per cent of the heating load is covered directly using electricity. On these cold days an air-water heat pump working with a flow temperature of 35 °C has a COP of around 2.5, we should note that heating with auxiliary electrical heating consumes 2.5 times as much electricity.

Now we examine the additional electricity costs for generating just that extra 3 per cent of additional heating:

The 97 per cent covered by the heat pump remains unchanged (i.e. W_{HP}). We multiply the 3 per cent for the auxiliary heating by the COP of 2.5, giving us 7.5 per cent. The total is then 97 per cent + 7.5 per cent = 104.5 per cent. That means that the cost of heating = electricity costs goes up by almost 5 per cent! That is a lot!

IMPORTANT

Even when the relationship W_{HP} / $W_{overall}$ initially looks good at 97 per cent, the extra costs for auxiliary electrical heating are considerable.

The auxiliary electrical heating consumes around 2.5 times as much electricity as an air-water heat pump!

It would make no sense to install an even smaller heat pump. Do the sums. It is never advisable to install an air-water heat pump that is too small. For example, where the heat pump covers only 80 per cent of the heating load, then the electricity costs go up by around 80 per cent + 20 per cent × 2.5 = 130 per cent. This no longer makes any sense.

5.6 Comparing water-water heat pumps, brine-water heat pumps and air-water heat pumps

The Table 5.16 shows a few advantages and disadvantages of the three types of heat pump:

This table is only a guide and depends upon the user's particular requirements. For example, if free cooling is very important to the user, then the two '++' are important. Where cooling is less important, then '+' or even '0' is sufficient.

5.7 Indicators of heat pump quality

The following quality indicators should be taken into account when selecting the best heat pump for the building in question and the individual needs of the operator:

1   How loud is the heat pump?
    You should not be able to hear the heat pump throughout the house, especially in houses without a basement! Look for good sound insulation in the heat pump housing, and several layers of acoustic insulation for the compressor.
2   Can the controller cool the heat pump?
    Can a cooling function be added later?
    This depends on the customer's particular requirements.

Table 5.16 Comparing the various types of heat pump

Water-water heat pumps

++	Best operational efficiency → Lowest operating costs
+	Inexpensive connection to heat source (well)
++	Includes well for garden irrigation, etc.
+	Heat pump inexpensive
−	Requires extensive planning
−	Well has limited lifespan
++	Very good free cooling possible
+	Monovalent operation possible
+	Limited space requirements
−	Some danger of freezing
+	Usually very quiet when operating

Brine-water heat pumps

+	Good operational efficiency → Low operating costs
−	High cost of borehole heat exchangers
++	Heat pump inexpensive
−	Requires extensive planning
++	Borehole or ground heat exchangers have very long lifespans
+	Free cooling with borehole heat exchangers
−	No free cooling with ground heat exchangers
+	Monovalent operation possible
+	Limited space requirements
++	Very high degree of operational safety
+	Usually very quiet when operating

Air-water heat pumps

−	Lower operational efficiency → Higher operating costs
++	No costs for connecting to heat source
−	Heat pump expensive
+	Needs little planning
−	No free cooling possible
−	No monovalent operation possible
−	Requires lots of space
+	High level of operational safety
−	Noisy during defrosting, especially for split units

3   Is the controller easy to use?
    The user interface on the controller should be simple, allowing the customer to make adjustments to the following functions:
    - Shifting the heating curve (e.g. with simple ± 3 °C adjustments).
    - Setting the required hot water temperature.
    - Holiday and party functions.
    All other settings, especially safety settings, should only be accessible to the specialist.
4   Does it have an integrated electrical heating element?
    This should not be the case, because then the customer cannot directly determine the level of electricity consumption. Electrical heating elements that are switched on by the controller should also be avoided.
5   Is the heat pump output sufficient to cover hot water generation as well as heating?
    For both ecological and economic reasons, the heat pump should provide both heating and hot water. It makes no sense to generate domestic hot water using an electrical heating element.
6   Does the controller have an interface for connection to a modem?
    This enables remote diagnosis if there is a malfunction.

5.8 Planning the electrical installation for a heat pump

Thought must also be given to planning the electrical installation. The following aspects must be considered:

1   If the customer wishes to take advantage of the special tariffs offered by many utilities, the savings this brings need to be weighed against the additional costs of a larger meter board, including additional meters, meter terminals, etc. This is not always worthwhile for smaller heat pumps. Additional meters usually incur extra meter fees.
2   The cut-out switch for the heat pump's alternating current must be of sufficient dimensions: the higher starting currents for the compressor must be considered when planning heat pump systems without soft starters.
3   In addition to the alternating current, there also needs to be a cut-out switch for the control voltage feed. During utility blocking times, this control voltage will drive the heat pump(s) and circulation pumps, etc., so that heating can continue using the buffer tank.

Attention must also be paid to local utility regulations.

For outdoor construction, the electrical connections are often rather long. High starting currents generate a large drop in voltage over the long cables. The utility needs to be consulted to determine whether the length of the cables and the starting current for the planned heat pump are suitable. This is especially the case for larger heat pumps, where outputs and starting currents are higher.

5.9 Applications and authorisations for heat pump systems

First, we need to differentiate between heat pump systems that require official permission and those that do not. Only air-water heat pumps do not require authorisation. Both water-water and brine-water heat pumps need official permission, usually from the local water authorities. In Germany at least, the local mining authorities must be given two weeks' notice before a well engineer drills over 100 m vertically into the earth. Ground and borehole heat exchangers and wells are not permitted in German water protection zones surrounding drinking water extraction areas. However, where drilling does not enter drinking water aquifers, it can be assumed that permission to drill borehole heat exchangers and install ground heat exchangers will be granted.

In Germany, applications for permission from water-related authorities (usually the local water board) and environmental bodies need to be supported by the following documentation:

- General map of the area, 1:25,000.
- Property plans, 1:1,000.
- Site plan, 1:500, indicating the location of the well and/or borehole or ground heat exchangers.
- An explanation of the planned measures.
- Purpose, including description of the system and its operation (brochures), calculation of the required lengths and dimensions of the ground heat exchangers, details on the geological and hydrological conditions, a description of the drilling procedures and sealing of the drilled holes.
- Supporting documentation and explanations, including the specialist documentation provided by the responsible drilling equipment operator, and sealing of the drilled holes in ground layers that separate aquifers.
- Details of the geological and hydrological characteristics of the subsoil.

Each local authority will provide information about the documentation that needs to be submitted.

In Germany, information about authorisation and the authorisation procedure is given in information sheets published by local state environment offices, and include the relevant regulations. These are, first and foremost, regulations that the well engineer must adhere to when drilling borehole heat exchangers for a brine-water heat pump system, or intake and discharge wells for a water-water heat pump system. They involve the so-called overriding general interests protecting aquifers. No dangerous substances must be permitted to enter the groundwater. The following rules must be observed:

- When drilling intake and discharge wells, the extracted water must be returned, after cooling, unchanged, to the same aquifer from which it came.
- Robust materials must be used when drilling the borehole heat exchangers, usually PE-HD pipes. No water-contaminating substances may be used when drilling. When inserting the borehole heat exchangers, the various aquifers must be kept separate from one another. To ensure smooth operation, the borehole heat exchangers must be compacted to a pressure of at least 15 bar before being filled with a glycol mixture. In water protection zones, there may be other prescribed mixtures for the brine, or permission for operating heat pumps denied. The well engineers should be certified (in Germany, DVGW Code of Practice W120) to demonstrate sufficient specialist knowledge in this area. In addition, there may be guidelines for drilling borehole heat exchangers (in Germany, VDI guideline 4640).
- When filling the borehole heat exchangers, filling connectors must be installed in the incoming and outgoing brine flows. These connectors are used to fill, rinse and vent the borehole heat exchangers. These connectors are subsequently closed and sealed,

to avoid any unauthorised filling or emptying. The brine pipes must be clearly distinguishable from all other piping.

After drilling, in accordance with quality assurance specifications, the drilling company must attest that the provisions regarding water law permits have been adhered to, and issue a test certificate for the borehole or ground heat exchangers.

• A monitoring system must be in place during heat pump operation to ensure that the heat pump is automatically shut down in the case of a leakage. This requires a pressure monitor for the closed brine circuit. A pressure safety valve with a response pressure of 10 bar must also be installed to protect against excess pressure.

If, despite all precautions, there is a leak in the borehole or ground heat exchangers, then the relevant authorities must be informed immediately.

Heat pumps must be decommissioned by a specialist. The heat transfer medium must be pumped out and the borehole heat exchangers, ground heat exchangers or wells compacted and closed using suitable materials.

Furthermore, in Germany, the following regulations and guidelines must be observed:

• LBO (state building codes)
  The LBO regulates the building code in each German state.
• EnEV (energy-saving ordinance)
  The EnEV regulates compliance with energy consumption limits designed to reduce CO_2 emissions.
• LwG (state water law)
  Each German state has its own water law to regulate groundwater use and protect groundwater quality.
• EEWärmeG (Renewable Energy Heat Act)
  38 per cent of energy consumption goes towards heating residential buildings. The EEWärmeG regulates the use of systems powered by renewables in order to reduce this level of energy consumption and/or encourage the use of renewables.

5.10 Funding for heat pump systems

In Germany, heating systems operated using renewables qualify for funding at federal, state or even municipal level, as well as, in some cases, from the utilities themselves. This also applies for heat pumps. Funding regulations are constantly changing and therefore we do not go into them in any more detail here. See information published online by each funding body.

In 2008 and 2009, heat pump systems generally qualified for funding when installed in new builds and existing building stock where it could be demonstrated that a minimum level of efficiency could be achieved. Unfortunately, this funding was discontinued in 2010, much to the surprise of many new heat pump operators who had been counting on this financial support. The funding was reinstated in the same year, although under much less favourable conditions. In principle, it excludes all new builds because the federal government already prescribes the use of renewables. Funding is only available for highly efficient heat pump systems in existing building stock, but this is unrealistic as the required framework conditions often do not exist. Highly efficient heat pump systems need low flow temperatures and these are rarely present in existing building stock, which is usually heated via radiators with high flow temperatures. In more favourable conditions, there may be underfloor heating on a ground floor, with radiators only on upper floors. Yet here too, the underfloor heating in existing buildings is often operated with significantly higher flow temperatures than those required for heat pump systems to work very efficiently.

5.11 The importance of surface heating systems for heat pumps

Heat pump systems can be operated most efficiently where the temperature of the heat source is as high as possible, and the heat sink (flow temperature) as low as possible. New builds should be planned with compact surface heating. Most surface heating is installed as underfloor heating, although it can also be installed in the form of wall surface or ceiling heating.

TIP

For heat pump systems using a buffer tank, the surface heating units must be installed using absolutely oxygen-tight pipes in order to avoid silting up as a result of corrosion.

---

6 Groundwater protection

Heat pumps play a significant role in protecting the environment. Both brine-water and water-water heat pumps use ambient heat from the ground (geothermal energy). Despite all precautions, leaks caused by material defects, installation errors or ageing cannot be ruled out, allowing brine to enter the soil. Leaks can also result in brine or other substances entering the groundwater. Borehole heat exchangers, in particular, are often installed at great depths, cutting through several aquifers, some of which are often sources of drinking water. Legally, it is not permitted to pollute groundwater, and, in Germany, this possibility must be excluded through the use of suitable measures as set down in LAWA 2002 (regional water authority specifications).

The welcome increase in the use of heat pumps and the steady ageing of existing installations also increases the likelihood of leaks. A water-glycol mixture is usually used as the heat transfer medium, and this must not be permitted to enter the groundwater.

6.1 Statutory provisions

Groundwater is a significant and valuable public resource, one that must be protected. There are many statutory provisions regulating the protection of groundwater by which manufacturers, installers and operators of heat pumps must abide.

Glycol is a German class 1 water hazard (WGK 1). The term WGK is defined in Germany's administrative regulations for materials hazardous to water (VwVwS).

According to § 62 of Germany's WHG (Water Resources Act), responsibility lies with both operators and builders. The builder must observe the technological regulations and the operator must ensure the system operates correctly. Both builder and operator are responsible for fulfilling § 3 (basic requirements) of the ordinance on industrial installations (VAwS). Systems must be built and operated in a manner that ensures that no hazardous substances can leak out. Care must be taken that possible leaks from any part of the system in contact with substances hazardous to water, including the borehole and ground heat exchangers, can be quickly and reliably detected. Where a leak does occur, the escaping substances must be quickly and reliably retained, and recycled or disposed of properly. In addition, operating instructions must be drawn up and adhered to, including a monitoring, maintenance and emergency plan. This is the responsibility of the manufacturer, constructor and operator.

It was not possible to fulfil these basic legal provisions with traditional engineering practice. This has now changed. Examination of the operating behaviour of the Geo-Protector with borehole heat exchangers undertaken by Prof. Dr. Stefan Wohnlich on behalf of the German Geological Society at the Ruhr University in Bochum clearly confirms this:

They show that for such leakages as are expected in practice, up to 100 per cent of which could escape, unhindered, using previous technological systems, the new Geo-Protector groundwater protection system can reduce this leakage to a minimum of between 0.05 per cent and 3.2 per cent. This means that the legal regulations can now be fulfilled.

TIP

This is now understood, thanks to a variety of publications as well as lectures and presentations of both the study results and the Geo-Protector. New criteria have now been established for modern engineering practice that enable the legal regulations to be almost completely fulfilled.

There are also positive economic benefits, as we will now examine.

6.2 Geo-Protector

The former German guideline VDI 4640 Part 2: 2001-09 and standard DIN 8901 required only that the heat pump is shut down and an alarm sounded when the brine pressure falls below a minimum level. However, this cannot prevent further brine escaping, as shown in Figure 6.1.

This means that in the case of an accident, the Geo-Protector as described below can limit the volume of escaping brine to an absolute minimum.

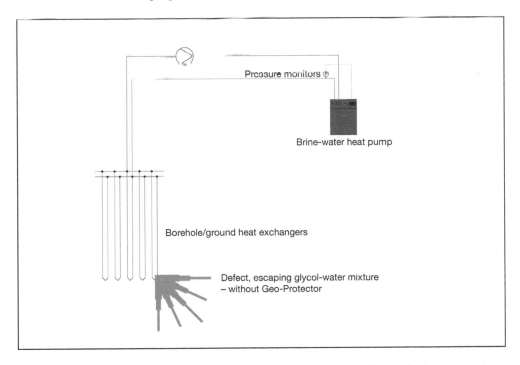

Figure 6.1 Escaping glycol-water mixture after the pressure switch shuts down the heat pump, in accordance with DIN 8901

The Geo-Protector, invented and patented by the author of this book, works as shown below.

6.2.1 The potential hazards associated with brine-water heat pumps

A brine-water heat pump system involves the following potential hazards:

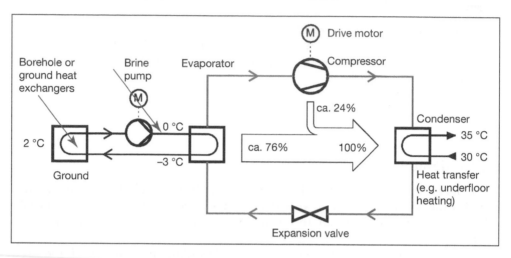

Figure 6.2 Potential hazards of a brine-water heat pump system

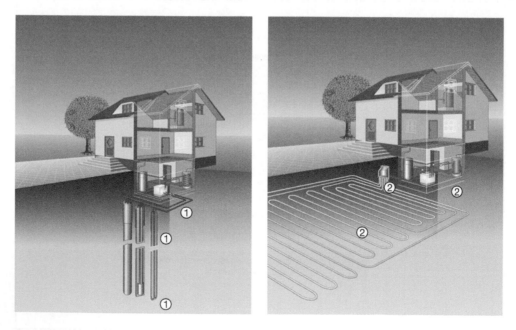

Figure 6.3 Potential hazards of a brine-water heat pump system with borehole and ground heat exchangers

1 Brine can escape along the entire length of the brine loops, contaminating the surrounding ground.

2 Here, too, an accident can lead to brine escaping and contaminating the soil.

6.2.2 The function of the Geo-Protector

The Geo-Protector essentially consists of at least two pressure monitors, three solenoid valves and a controller.

The Geo-Protector uses the two pressure monitors to constantly oversee the brine pressure. When there is a leak, brine escapes from the brine circuit. The pressure falls. If the brine pressure falls below the lowest permitted pressure P_{1min} over a longer period, then the Geo-Protector warns the operator by means of a pre-alarm. At this point, the operator has the option of checking the heat pump system's brine circuit; the low pressure could be caused by the brine cooling down, and in this case the expensive brine does not automatically need to be rinsed out. To ensure that operational pressure fluctuations do not result in a shutdown, the pressure P_{1min} must be undercut for a duration of more than 5 minutes, for example (the length of time is adjustable).

Where there is a further, persistent fall in pressure below P_{2min}, then the main alarm will be triggered, again after a preset delay period. Again, to ensure that operational pressure fluctuations do not automatically lead to the system being shut down, the pressure P_{2min} must be undercut for a longer duration than, for example, 5 minutes (this time is adjustable) before the main alarm is sounded. This prevents operational pressure fluctuations causing the brine to be rinsed out, and the heat pump shut down.

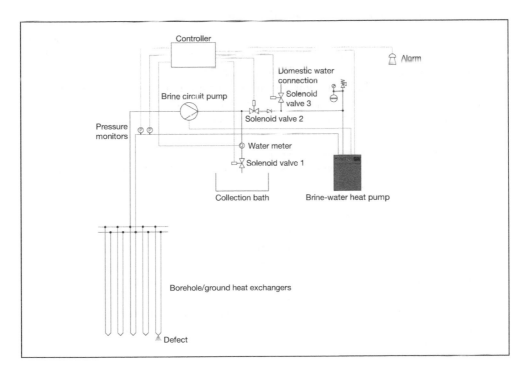

Figure 6.4 Function of the Geo-Protector

When the main alarm is sounded, rinsing begins and the heat pump stays shut down. Brine-water heat pumps are the most common type of heat pump although they are divided into those using borehole or ground heat exchangers, drainage collectors (ditch collectors), energy baskets and others. However, they all have one thing in common – damage can cause a leak in the brine circuit.

In order to rinse out the brine circuit with drinking water, the solenoid valve 2 closes and simultaneously the solenoid valves 1 and 3 open for a preset period. The length of the rinsing period depends on the size of the heat pump and/or brine circuit, and this can vary significantly depending on the type of heat pump. The brine pump stays turned on during rinsing.

Usually, the brine circuit in a brine-water heat pump system consists of several loops, all attached to a circuit manifold. Two pipes run from the manifold to the heat pump (incoming and returning flow). Consequently, the brine circuit manifold itself is a further potential source of leakage. Therefore, the brine circuit manifold must be installed inside a sealed distribution shaft. A further sensor (float switch, conductivity meter, etc.) can also be installed in the shaft, setting off an alarm if there is a leak in the shaft. This would then turn off the heat pump without instigating the extraction and rinsing process – because the distribution shaft itself would then be flooded, allowing brine to escape into the environment.

The Geo-Protector is a relatively inexpensive means of significantly increasing the level of environmental protection, making a heat pump even more environmentally friendly. The system can be used for all types of heat pump (i.e. for water-water heat pumps with an intermediate circuit for system separation, or for brine-water heat pumps for borehole and ground heat exchangers, as well as for direct evaporating heat pump systems).

6.2.3 Environmental and economic considerations

Groundwater is a valuable public resource that must be protected in accordance with all the legal regulations. Previously, the German guideline VDI 4640 Part 2: 2001-09 was applicable for regulating groundwater protection, directing that once pressure had fallen below a minimum level, the heat pump system be shut down and an alarm triggered.
The additional costs involved in implementing this former regulation were that of a pressure monitor and a one-time modification to the controller.

Today's technical state of the art is greater groundwater protection thanks to the Geo-Protector or Geo-Protector system. When pressure falls, a pre-alarm is triggered and the heat pump is shut down. Only when pressure continues to fall is the brine circuit rinsed out with drinking water. Then, no more brine can leak out, only drinking water. *The examinations undertaken by the Ruhr University show that when the Geo-Protector is used, only 0.05 per cent to 3.2 per cent of the brine can escape once rinsing has started.*

The additional costs for this level of protection include a further pressure monitor, three small solenoid valves and a one-time modification to the controller. For a simple heat pump system with borehole heat exchangers and integration into the heat pump controller, this represents only around 1 per cent of the total cost – an amount that can certainly be recovered when discounts are negotiated.

So it is now possible to significantly increase groundwater protection at minimal cost, and, where there is a leak, to significantly reduce the resultant damage – from virtually 100 per cent to around 2 per cent.

Therefore, it is inconsistent and negligent not to spend the minimal additional costs needed to provide this protection.

6.2.4 Cost-benefit considerations

We now look at the costs and benefits of the former and current levels of groundwater protection technology. There are several reasons for integrating groundwater protection in the heat pump controller. They include the need to protect groundwater, a valuable public resource. In addition, by making integration of groundwater protection into the controller standard, the unit cost for controllers falls significantly, and the need to purchase Geo-Protector hardware is avoided. See Figure 6.4, function of the Geo-Protector.

Former state of the art:

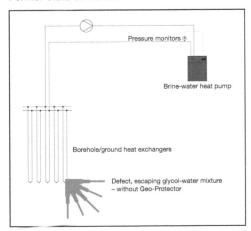

Current state of the art:

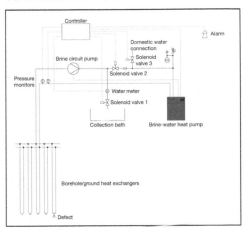

Costs:
- one pressure monitor
- one software adaptation – one-time cost

Costs:
- two pressure monitors
- one non-return valve
- two small solenoid valves
- one software adaptation – one-time cost

Integrating groundwater protection into the controller involves additional costs for an extra pressure monitor, a non-return valve and two small solenoid valves – probably all covered by the negotiated discounts available when purchasing a heat pump. The impact on the customer's wallet is negligible.

This is a means of making environmentally friendly heat pump technology even more desirable, and at minimal cost.

6.2.5 Recommendations for implementing a Geo-Protector

The following points should be considered when using a Geo-Protector in order to ensure safe operation and optimal groundwater protection:

When dimensioning the brine expansion tank, it is vital that this is no larger than necessary: the larger the tank, the more brine that will be expelled in the event of a leak. A brine expansion tank should be dimensioned according to DIN EN 12828.

For cost reasons, the Geo-Protector system should be integrated into the heat pump controller right from the start. This involves one-time software programming costs. Using an external device involves additional hardware costs for the monitoring device.

When the pressure falls below the first low pressure level (pre-alarm), an alarm is triggered and the heat pump is shut down. If the pressure falls below the second low pressure level (main alarm), then the rinsing procedure starts and the heat pump remains shut down. A locking mechanism ensures that the heat pump is prevented from restarting. Where the pressure repeatedly falls below the first low pressure level (pre-alarm) within a relatively short time period, then rinsing should also start (main alarm).

6.3 CO_2 borehole heat exchangers

CO_2 borehole heat exchangers are filled with CO_2. They are comparable to the heat pipe technology in solar technology. Geothermal energy causes the CO_2 gas in the borehole heat exchanger to evaporate. The CO_2 gas rises and condenses on the heat exchanger between borehole heat exchanger and brine circuit. The condensation heat is transferred to the heat pump circuit. The condensed CO_2 is now a liquid that flows back down the walls of the borehole heat exchanger until it evaporates again. Here again, large quantities of heat (energy) are released as the gas changes state.

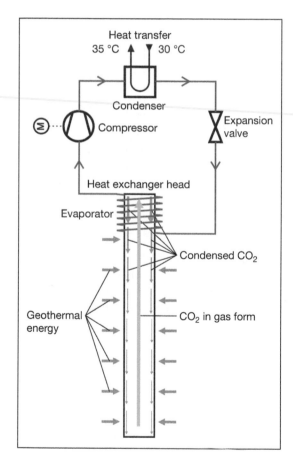

Figure 6.5 CO_2 borehole heat exchanger

The heat transferred from the CO_2 borehole heat exchangers can be transferred via a brine circuit or, as shown below, directly to the heat pump.

CO_2 borehole heat exchangers have a particular environmental benefit: the borehole heat exchangers contain only a totally environmentally friendly gas. Should there be a leak, the groundwater contamination, if any, is minimal, because the gas rises. If it comes into contact with water, only a small portion will be absorbed, and this is also determined by the nature of the water itself.

Consequently, a heat pump that operates using CO_2 borehole heat exchangers is a major contribution to environmental protection.

CO_2 borehole heat exchangers can be compared to the raw materials of a direct evaporator. Instead of cheaper PE pipes, gas-tight and therefore more expensive pipes must be installed.

7 Hydraulics

A good hydraulic balance is vital if a heating system is to work properly and efficiently, especially a heat pump system. The choice of primary and secondary pump is particularly important: having the right pump ensures the flow volume needed for the heating system to operate optimally.

The following diagram shows a heat pump system:

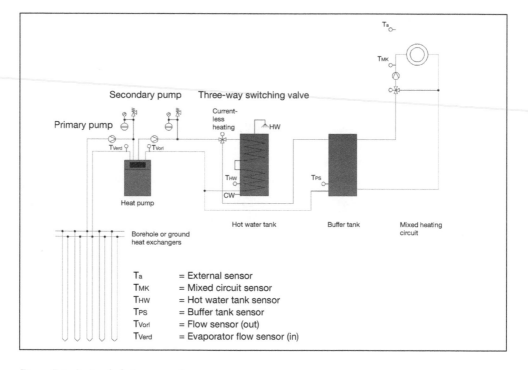

T_a	= External sensor
T_{MK}	= Mixed circuit sensor
T_{HW}	= Hot water tank sensor
T_{PS}	= Buffer tank sensor
T_{Vorl}	= Flow sensor (out)
T_{Verd}	= Evaporator flow sensor (in)

Figure 7.1 A simple brine-water heat pump system

A heat pump system consists of the following key components:

1   heat source, here borehole or ground heat exchangers, and, for water-water heat pumps, intake and discharge wells;
2   primary pump, here the brine pump, and, for water-water heat pumps, an underwater heat pump;

3   heat pump;
4   secondary pump or charging pump;
5   three-way switching valve;
6   hot water tank;
7   buffer tank; and
8   heating circuit.

7.1 The primary pump

Primary pumps include the pumps that supply the heat pump with the 'heat' from the environment. For a water-water heat pump, this is the water pump, usually an underwater pump; for a brine-water heat pump, the brine circulation pump; and for an air heat pump, it is the fan/ventilator.

Underwater pumps are normally used for water-water heat pumps. These are electric pumps installed in the intake well of a water-water heat pump system. Alternatively, a centrifugal pump can be installed above ground.

It is very important that the selected pumps are powerful enough, and the technical data provided by the manufacturer must be observed. Generally, we can assume that 10 kW is needed for a minimum flow of 2 m³/h, or 20 kW for roughly 4 m³/h.

Where the flow rate is less than the given minimum, the heat transfer medium can cool excessively, damaging the heat pump or causing it to be shut down.

In a water-water heat pump, the heat transfer medium is water.

PLEASE NOTE

Excessive cooling can lead to freezing!
Therefore, it is vital that the primary pump for a water-water heat pump, usually an underwater pump, is not undersized.
Remember that the water level in the well falls as the pump operates.

7.1.1 The underwater pump and how it is monitored

An underwater pump is usually the primary pump for a water-water heat pump.

The disadvantage of centrifugal pumps, which are not installed inside the well, is that if the water column is broken, then the heat pump is no longer supplied with water.

When operating a water-water heat pump, it is vital that the evaporator is fed with at least the minimum flow of well water. If the flow falls below this minimum rate, the evaporator can freeze. This often leads to the complete shutdown of a heat pump and involves expensive repairs. Therefore, the primary pump must be sufficiently powerful. If it is too powerful, however, then its energy consumption will be too high.

In order to determine the correct size of pump, we first need to establish the minimum flow rate in accordance with the technical details provided by the heat pump manufacturer. We also need to determine the overall pressure loss for the system. The following equation is used to determine total pressure loss:

$$\Delta P = P_P + P_{MB} + P_{HPE} + P_H$$

Table 7.1 Flow speeds and pressure loss across PE pipes

Pressure loss across PE plastic pipes

Plastic pipes PELM/PEW PEN 10

Flow volume			PELM				PEW							
m³/h	l/min	l/sec	25 / 20.4	32 / 26.2	40 / 32.6	50 / 40.6	63 / 51.4	75 / 61.4	90 / 73.5	110 / 90.0	125 / 102.2	140 / 114.6	160 / 130.8	180 / 147.2
0.6	10	0.16	0.49 / 1.8	0.30 / 0.66	0.19 / 0.27	0.12 / 0.085								
0.9	15	0.25	0.76 / 4.0	0.46 / 1.14	0.3 / 0.6	0.19 / 0.18	0.12 / 0.63							
1.2	20	0.33	1.0 / 6.4	0.61 / 2.2	0.39 / 0.9	0.25 / 0.28	0.16 / 0.11							
1.5	25	0.42	1.3 / 10.0	0.78 / 3.5	0.5 / 1.4	0.32 / 0.43	0.2 / 0.17	0.14 / 0.074						
1.8	30	0.50	1.53 / 13.0	0.93 / 4.6	0.6 / 1.9	0.38 / 0.57	0.24 / 0.22	0.17 / 0.092						
2.1	35	0.58	1.77 / 16.0	1.08 / 6.0	0.69 / 2.0	0.44 / 0.70	0.28 / 0.27	0.2 / 0.12						
2.4	40	0.67	2.05 / 22.0	1.24 / 7.5	0.80 / 3.30	0.51 / 0.93	0.32 / 0.35	0.23 / 0.16	0.16 / 0.063					
3.0	50	0.83	2.54 / 37.0	1.54 / 11.0	0.99 / 4.8	0.63 / 1.40	0.4 / 0.50	0.28 / 0.22	0.2 / 0.09					
3.6	60	1.00	3.06 / 43.0	1.85 / 15.0	1.2 / 6.5	0.76 / 1.90	0.48 / 0.70	0.34 / 0.32	0.24 / 0.13	0.16 / 0.050				
4.2	70	1.12	3.43 / 50.0	2.08 / 18.0	1.34 / 8.0	0.86 / 2.50	0.54 / 0.83	0.38 / 0.38	0.26 / 0.17	0.18 / 0.068				
4.8	80	1.33		2.47 / 25.0	1.59 / 10.5	1.02 / 3.00	0.64 / 1.20	0.45 / 0.50	0.31 / 0.22	0.2 / 0.084				
5.4	90	1.5		2.78 / 30.0	1.8 / 12.0	1.15 / 3.50	0.72 / 1.30	0.51 / 0.57	0.35 / 0.26	0.24 / 0.092	0.18 / 0.05			
6.0	100	1.67		3.1 / 39.0	2.0 / 16.0	1.28 / 4.6	0.8 / 1.8	0.56 / 0.73	0.39 / 0.30	0.26 / 0.17	0.2 / 0.07			
7.5	125	2.08		3.86 / 50.0	2.49 / 24.0	1.59 / 6.6	1.00 / 2.50	0.70 / 1.10	0.49 / 0.50	0.33 / 0.18	0.25 / 0.10	0.20 / 0.55		
9.0	150	2.50		3.0 / 33.0		1.91 / 8.6	1.20 / 3.5	0.84 / 1.40	0.59 / 0.63	0.39 / 0.24	0.30 / 0.13	0.24 / 0.075		

10.5	175	2.92		3.5	2.23	1.41	0.99	0.69	0.46	0.36	0.28	
				38.0	11.0	4.3	1.80	0.78	0.30	0.18	0.09	
12.0	200	3.33		3.99	2.55	1.6	1.12	0.78	0.52	0.41	0.32	0.25
				50.0	14.0	5.5	2.4	1.0	0.40	0.22	0.12	0.065
15	150	4.17			3.19	2.01	1.41	0.98	0.66	0.51	0.40	0.31	0.25
					21.0	8.0	3.70	1.50	0.57	0.34	0.18	0.105	0.06
18	300	5.00			3.82	2.41	1.69	1.18	0.78	0.61	0.48	0.37	0.29
					28.0	10.5	4.60	1.95	0.77	0.45	0.25	0.13	0.085
24	400	6.67				3.21	2.25	1.57	1.05	0.81	0.65	0.50	0.39
						19.0	8.0	3.60	1.40	0.78	0.44	0.23	0.15
30	500	8.3				4.01	2.81	1.96	1.31	1.02	0.81	0.62	0.49
						28.0	11.5	5.0	2.0	1.20	0.63	0.33	0.21
36	600	10.0				4.82	3.38	2.35	1.57	1.22	0.97	0.74	0.59
						37.0	15.0	6.6	2.6	1.50	0.82	0.45	0.28
42	700	11.7				5.64	3.95	2.75	1.84	1.43	1.13	0.87	0.69
						47.0	24.0	8.0	3.5	1.90	1.10	0.60	0.40
48	800	13.3					4.49	3.13	2.09	1.62	1.29	0.99	0.78
							26.0	11.0	4.5	2.60	1.40	0.81	0.48
54	900	15.0					5.07	3.53	2.36	1.83	1.45	1.12	0.80
							33.0	13.5	5.5	3.20	1.70	0.95	0.58
60	1000	16.7					5.64	3.93	2.63	2.04	1.62	1.24	0.96
							40.0	16.0	6.7	3.90	2.2	1.2	0.75
75	1250	20.8						4.89	3.27	2.54	2.02	1.55	1.22
								25.0	9.0	5.0	3.0	1.6	0.95
90	1500	25.0						5.88	3.93	3.05	2.42	1.86	1.47
								33.0	13.0	8.0	4.1	2.3	1.40
105	1750	29.2						6.86	4.59	3.56	2.83	2.17	1.72
								44.0	17.5	9.7	5.7	3.2	1.9
120	2000	33.3							5.23	4.06	3.23	2.48	1.96
									23.0	13.0	7.0	4.0	2.4
150	2500	41.7							6.55	5.08	4.04	3.10	2.45
									34.0	18.0	10.5	6.0	3.5
180	3000	50							7.86	6.1	4.85	3.72	2.94
									45.0	27.0	14.0	7.6	4.0
240	4000	66.7								8.13	6.47	4.96	3.92
										43.0	24.0	13.0	7.5
300	5000	83.3								8.08	6.2	4.89
										33.0	18.0	11.0

The upper figures give the flow speed in m/sec; the lower figures give the pressure loss per 100 m of straight pipe in m. Roughness: K = 0.01 mm, T = 10°C

where P_P = pressure loss across the pipes (incoming and outgoing) [bar], [kPa]
$\quad\quad$ P_{MB} = pressure loss across all moulded parts and bends [bar], [kPa]
$\quad\quad$ P_{HPE} = pressure loss across the heat pump evaporator – see technical details for each heat pump [bar], [kPa]
$\quad\quad$ P_H = pressure loss dependent on the height of the water table + sinking in the well [bar]

TIP

If an additional heat exchanger is being used for free cooling, then the pressure loss across this additional exchanger must also be taken into account.

Table 7.1 can be used to determine the pressure loss across the feed and return PE pipes.

TIP

- Pressure loss across bends, moulded and T-shaped parts and fittings must also be taken into account.

Table 7.2 Pressure loss across bends, T-shaped parts and fittings

Nominal width	1″	1¼″	1½″	2″	2½″	3″	3½″	4″	5″	6″
90° bends and shut-off valves	1.1	1.2	1.3	1.4	1.5	1.6	1.6	1.7	2.0	2.5
T-shaped parts and non-return valves	4.0	5.0	5.0	5.0	6.0	6.0	6.0	7.0	8.0	9.0

The pressure loss across bends, gate valves, T-shapes and non-return valves are the equivalent to the length of a straight pipe, as shown in the last two rows of the table.

TIP

If using longer pipe lengths, it may be sensible to choose slightly larger pipes.

The following apply when planning the piping:

- The flow speed through the pipes should not exceed a maximum of 1.5 m/sec, preferably 1.0 m/sec.
- Pressure losses mean energy losses.
  Therefore pipes should be too large rather than too small.
- The length of the pipes should also be considered.

- Pressure loss across bends, T-shapes and fittings should not be forgotten.
- The well engineer can provide information about pressure loss caused by the height of the water table. Calculations must start from the lowest expected water table level, plus the fall of the water level in the intake well as a result of ongoing operation.
- An adjustable flow monitor is recommended as a means of protecting a water-water heat pump. Without a flow monitor, the warranty on the evaporator (heat exchanger on the water side) is no longer valid.

Note:
The details given above are an aid to dimensioning the piping. The technical information provided by each heat pump manufacturer should form the basis for planning.

Once the overall pressure loss (feed + return pipes + connectors + heat pump) as well as the minimum required flow rates (see manufacturer's technical details) are known, then the correct underwater pump can be selected according to the following characteristics:

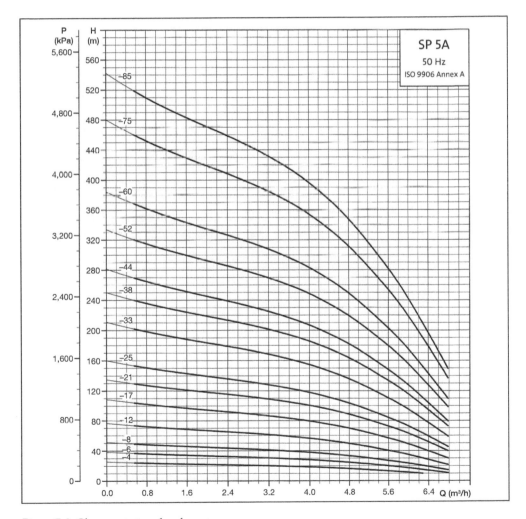

Figure 7.2 Characteristics of underwater pumps

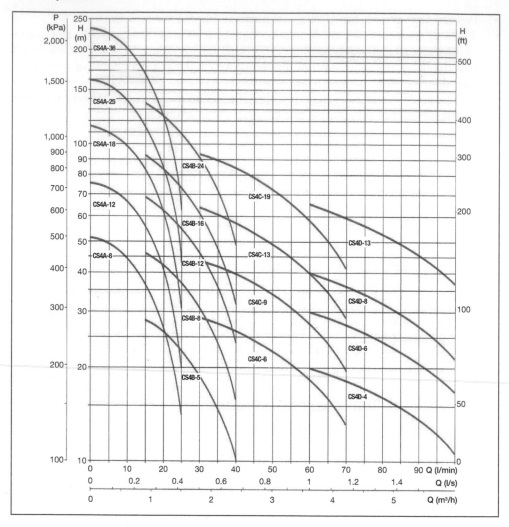

Figure 7.3 Characteristics of underwater pumps

The following monitoring devices are used to monitor the water supply:

1   the motor protection switch to protect the motor of the underwater pump;
2   a flow monitor to monitor the flow to the heat pump; and, if necessary,
3   dry running protection to protect both the underwater pump and the heat pump.

Motor protection switch
A motor protection switch is essential and should be integrated into the heat pump. It protects the underwater pump's electric motor, switching the pump off when overloaded. The auxiliary contact sends an alarm signal to the heat pump controller, which in turn immediately switches off the heat pump to protect the evaporator from freezing.

Flow monitor

A flow monitor is highly recommended for safe operation of a heat pump. The flow monitor must be set so that it immediately shuts down the heat pump when the flow rate falls below a set minimum. It is an additional freezing protection for the evaporator. The flow monitor should be set in accordance with the data provided by the manufacturer.

- *Correctly adjusting the flow monitor is essential for the heat pump's operational safety!*

Protection against dry running

Dry running protection is primarily designed to protect the underwater pump, although it also protects the heat pump. If the pump is in danger of running dry (destroying the motor by overheating), then the dry running protection turns off the underwater pump.

IMPORTANT

The dry running protection must be electrically linked to the heat pump so that the heat pump is also immediately shut down if the underwater pump is turned off.

7.1.2 Planning a water supply with underwater pump and piping

Let us stay with our sample project requiring a heating output of 10.64 kW, and needing a heat pump with a minimum output of 12.8 kW:

From the technical documentation, we see that the heat pump with an output of 14.2 kW needs to be supplied with a minimum flow rate of 2.4 m^3/h.

First, the geological conditions must be established. An intake and a discharge well with a minimum pumping capacity of around 5 m^3/h are required.

- When determining the well's minimum output, we first need to know the output of the planned underwater pump. The well must also have the necessary reserves to cope with its decreasing intake and absorption capacities over time.

Ideally, the water quality should also be investigated, by taking samples from neighbouring wells or asking an experienced well engineer. Here, we will assume that both outcomes are positive:

1 The well engineer submits an offer for drilling an intake and a discharge well.
2 The water is probably free of iron and manganese, or they are present at an acceptable level.

Now we need to choose the correct pump. Here, we use the technical data for underwater pumps. The minimum flow rate for this heat pump is 2.4 m^3/h.

Figure 7.6 shows that one of these underwater pumps is suitable.

The underwater pump SP 2A-6 has a delivery head of around 15 m at 2.4 m^3/h. This corresponds to a pressure of 1.5 bar. Now we have to check if this pressure is sufficient.

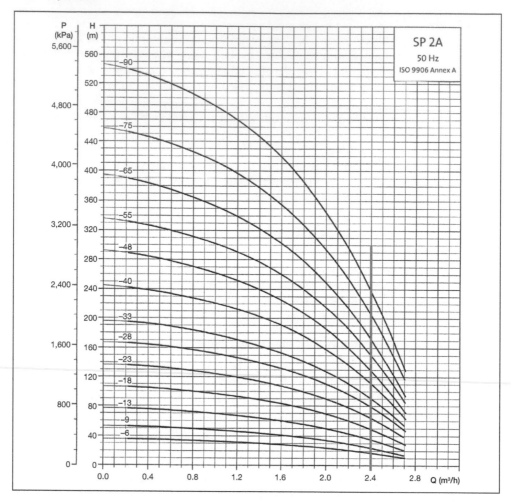

Figure 7.4 Characteristics of underwater pumps

Now we have to dimension the pipes:

For a medium-sized, detached house, the intake and discharge wells should be at least 15 m to 20 m apart. The wells should always be accessible. Therefore, the wells are planned as follows:

The wells are drilled correctly by a well engineer. As the planned underwater pump can pump up to 2.7 m³/h, both wells should have a permanent pumping capacity of at least 3 m³/h.

Overall length of the PE piping: approx. 40 m.

The table above for pressure drop shows that we need PE pipes with a minimum diameter of 32 mm, preferably 40 mm.

- Note that this measurement refers to the external diameter. Therefore, a DN 32 = 1″, and DN 40 = 1¼″.
- The larger the diameter, the lower the pressure loss and therefore power consumption of the underwater pump.

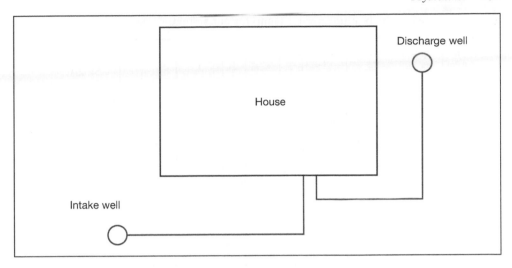

Figure 7.5 Intake and discharge well for a detached house

We choose the PE pipe DN 32. A pressure resistance of 6 bar is entirely adequate. Assuming a flow rate of 2.4 m³/h, the table for pressure drop in PE pipes gives us a pressure loss of:

$$P_P = 7.5 \text{ mWs}/100 \text{ m} \times 40 \text{ m} = 3.0 \text{ mWs} = 0.3 \text{ bar}$$

Now we need to calculate the pressure loss across the moulded parts and bends:

$$P_{MB} = 1.2 \text{ m} \times 7 \times 7.5 \text{ mWs}/100 \text{ m} = 0.63 \text{ mWs} - 0.06 \text{ bar}$$

The pressure loss across the heat pump evaporator is listed in the technical data issued by the heat pump manufacturer. Thus:

$$P_{HPE} = 0.1 \text{ bar}$$

The minimum expected groundwater level during operation, together with the fall in water column height in the intake well, is 15 m. This corresponds to an additional pressure loss of:

$$P_H = 15 \text{ mWs} = 1.5 \text{ bar}$$

And therefore the overall pressure loss is calculated at:

$$\Delta P = P_P + P_{MB} + P_{HP} + P_H$$
$$= 0.3 \text{ bar} + 0.06 \text{ bar} + 0.1 \text{ bar} + 1.5 \text{ bar}$$
$$= 1.96 \text{ bar}$$

To be on the safe side, the estimated pressure loss should not be too low. Therefore, here we will assume 2 bar = 20 mWs.

Now we can select the best underwater pump for our system:

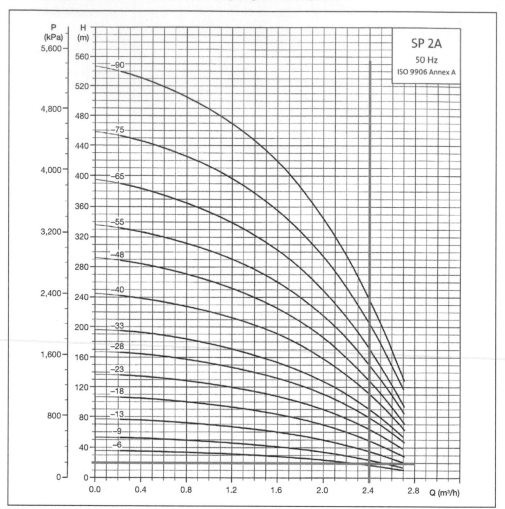

Figure 7.6 Selecting the right underwater pump

All underwater pumps whose characteristics fall below the intersecting lines are excluded. The smallest possible underwater pump is therefore model SP 2A-9.

As the characteristics of the underwater pump are significantly above the calculated operating point, we can assume that the flow rate will be correspondingly higher (e.g. 2.7 m³/h).

Now we have to choose a flow monitor that will turn off the heat pump when the minimum flow falls below 2.4 m³/h. In practice, when starting up a heat pump for the first time, the flow monitor is programmed to turn on, but not off, when the pump is running. But as we are selecting the next largest underwater pump (or centrifugal pump), the flow switch setting is adequate.

Now our planning for the water side of the water-water heat pump system is complete. The next task is to plan the secondary pump.

7.1.3 The brine circulation pump

The brine circulation pump circulates the brine through the borehole or ground heat exchangers.

Three points must be considered when dimensioning the brine circulation pump:

1 the minimum flow rate;
2 the total pressure loss across the cooling circuit; and
3 the heat transfer medium.

Refer to the manufacturer's data for the minimum flow rate. This minimum must be observed. However, the pump should not be too large. After dimensioning, you should choose the next brine circulation pump up in size.

The total pressure loss is calculated as the sum of the following components:

1 pressure loss across the brine circuit (borehole or ground heat exchangers);
2 pressure losses in the connecting pipes;
3 pressure loss across the moulded parts (manifold and bends); and
4 pressure loss across the heat pump evaporator.

The total pressure loss is calculated as:

$$\Delta P = P_B + P_P + P_{MB} + P_{HPE}$$

where
ΔP = total pressure loss across the brine circuit [bar]
P = pressure loss over a brine loop = pressure loss over all brine loops [bar], [kPa]
P_P = pressure loss across the incoming and outgoing pipes [bar]
P_{MB} = pressure loss across all moulded parts and bends [bar]
P_{HPE} = pressure loss across the heat pump evaporator – see the technical data provided by the heat pump manufacturer [bar], [kPa]

• Do not forget that double-U borehole heat exchangers each have two brine loops.

TIP

If the system includes free cooling using an additional heat exchanger, then remember to include the pressure loss across this heat exchanger.

The pressure loss over a brine loop can be determined using the table for calculating the pressure drop. In order to do this, the brine flow for each brine loop must first be calculated:

$$Q_B = Q_{HP} / n$$

where
Q_B = brine flow per loop [m³/h]
Q_{HP} = total flow = minimum flow rate through the heat pump [m³/h]
n = number of brine loops

Now we know the minimum flow rate and the pressure loss, we can choose the brine pump.

7.1.4 *Planning example for a brine circuit network with a brine circulation pump and pipes*

We will stay with our previous project example, with a required heat output of 12.4 kW: According to the manufacturer's data (GeoMax®), a brine-water heat pump with an output of 12.4 kW requires a brine circulation pump with 2.7 m³/h.
Now we need to calculate the pressure loss. The total pressure loss is calculated as:

$$\Delta P = P_B + P_P + P_{MB} + P_{HPE}$$

In our house project calculations, we determined that three borehole heat exchangers using PE pipes with DN = 25 and a depth of 75 m were needed. A further 2 × 10 m = 20 m are needed for connecting the borehole heat exchangers to the brine circuit manifold. Thus the total pipe length for a loop:

$$L_L = 2 \times (75\ m + 20\ m) = 170\ m$$

Three double-U pipe borehole heat exchangers have a total of n = 2 × 3 = 6 loops. When installed according to Tichelmann, the brine flow of Q = 2.7 m³/h is divided across six loops. Therefore, each borehole heat exchanger loop has a flow of:

$$Q_B = Q_T / n = 2.7\ m^3/h / 6 = 0.45\ m^3/h$$

Using the table for pressure drop for PE pipes, the pressure loss over a single brine loop can be determined. The pressure loss for a loop with a minimum flow of 0.6 m³/h, for a PE pipe with DN = 25:

$$P_B = 1.8\ mWs/100\ m \times 170\ m$$

$$= 2.95\ mWs$$

$$= 0.3\ bar\ (at\ 0.6\ m^3/h)$$

TIP

The pressure loss is certainly less with a flow rate of 0.45 m³/h.

The length of the pipes (feed and return) from the brine circuit manifold to the heat pump should be kept as short as possible. Here, they are both 10 m, giving us a total length of 20 m. A PE pipe with DN = 32 runs from the manifold to the heat pump. With a flow rate of 2.7 m3/h, the table for pressure loss across PE pipes gives us a pressure loss of:

$$P_P = 11\ mWs/100\ m \times 20\ m$$

$$= 2.2\ mWs$$

$$= 0.22\ bar\ (at\ 3.0\ m^3/h)$$

The pressure loss across the heat pump evaporator can be taken from the manufacturer's technical data. Here:

$$P_{HPE} = 0.3 \text{ bar}$$

The total pressure loss is now calculated as:

$$\Delta P = P_B + P_P + P_{MB} + P_{HPE}$$

$$= 0.30 \text{ bar} + 0.22 \text{ bar} + 0.1 \text{ bar} + 0.3 \text{ bar}$$

$$= 0.92 \text{ bar} = 9 \text{ mWs}$$

We use these values to select the brine circulation pump:

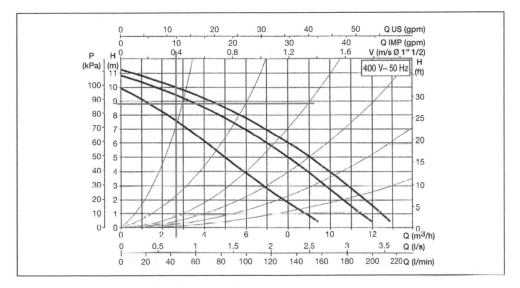

Figure 7.7 Choosing the brine circulation pump

The brine circulation pump with these characteristics is sufficient. It needs to be set at level 2.

Now we need to consider the heat transfer medium, usually a glycol-water mixture, because a glycol-water mixture has a different viscosity to water. Often the mixtures can contain up to 30 per cent glycol. Here again, attention must be paid to the manufacturer's specifications.

7.2 The secondary or charging pump

The manufacturer's specifications must also be considered when planning the charging pump(s).

First, we need to decide whether to use a charging pump and a switching valve, or two charging pumps. If using two charging pumps, it is important to ensure that only one can be turned on at a time. Usually, one charging pump is used in combination with a switching valve.

The minimum flow rates must be considered when selecting the charging pump(s), especially for hot water generation: at high temperatures and with too little heat transfer (which occurs when the heat pump is too small), the heat pump gets shut down if there is a high pressure fault.

Now we need to calculate the pressure losses. The pressure loss across the hot water tank is greater than the loss across the buffer tank. Therefore, we must take into account the pressure loss across the hot water tank.

The total pressure loss is calculated as:

$$\Delta P_H = P_{HPC} + P_{PH} + P_{SV} + P_{MBH} + P_T$$

where P_{HPC} = pressure loss over the condenser in the heat pump [bar], [kPa]
  P_{PH} = pressure loss over the piping on the heating side [bar]
  P_{SV} = pressure loss over the three-way switching valve [bar], [kPa]
  P_{MBH} = pressure losses over moulded parts, bends, T-shapes, etc. [bar]
  P_T = pressure loss across the heat exchanger in the hot water tank [bar], [mbar]

The manufacturer's specifications will list the pressure loss over the condenser in the heat pump P_{HPC}.

The precise pressure loss through the piping P_{PH} is listed by the manufacturer. We use the table for pressure drop over PE pipes again for an initial approximation, because the pressure losses in steel, copper or smooth-walled composite plastic pipes are somewhat less than in PE pipes. This keeps us on the safe (although less favourable) side.

Pressure loss over three-way switching valves must also be considered, and, here again, we need to refer to the manufacturer's specifications. A K_{VS} value is used for control elements. Once we know the flow rate, and with the aid of the K_{VS} value, we can determine the pressure loss, for example by referring to the following diagram:

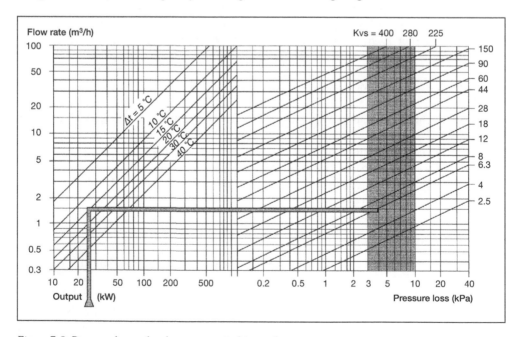

Figure 7.8 Pressure losses for three-way switching valves

Table 7.3 Pressure losses and technical data for various hot water tanks

		N_L-number	Throughput in kW or l/h															Flow resistance in mbar		
Flow temperature	Heated area	80°C	70°C	70°C	70°C	[illegible]°C	80°C	80°C	70°C	70°C	70°C	80°C	80°C	80°C			
Hot water temperature		45°C	45°C	45°C	45°C	[illegible]°C	45°C	45°C	60°C	60°C	60°C	60°C	60°C	60°C			
Cold water temperature		10°C	10°C	10°C	10°C	[illegible]°C	10°C	10°C	10°C	10°C	10°C	10°C	10°C	10°C			
Thermostat setting		60°C															
Hot water flow rate	m²	3 m³/h	1 m³/h	2 m³/h	3 m³/h	1 m³/h	2 m³/h	3 m³/h	1 m³/h	2 m³/h	3 m³/h	1 m³/h	2 m³/h	3 m³/h	1 m³/h	2 m³/h	3 m³/h
BE 160 ERM	0,60	2	11,9	13,9	14,9	[illegible]	18,5	13,8	8,5	9,7	10,3	12,5	14,7	15,7	2	14	54
			293	342	367	[illegible]	455	487	146	167	177	215	253	270			
BE 200 ERM	1,00	3,5	18,0	21,7	23,5	[illegible]	28,4	31,0	13,2	15,5	16,6	19,1	22,9	24,8	14	45	96
			443	534	578	[illegible]	699	763	227	267	286	329	394	427			
BE 300 ERM	1,50	7,5	23,0	30,1	31,8	[illegible]	39,1	42,7	17,1	20,9	22,4	24,8	31,0	33,9	32	90	178
			566	740	782	[illegible]	962	1050	294	360	386	427	534	584			
BE 400 ERM	1,80	11	27,2	34,8	38,9	[illegible]	45,1	50,7	20,4	25,5	27,5	29,3	37,1	41,2	53	114	210
			669	856	957	[illegible]	1109	1247	351	439	474	505	639	709			
BE 500 ERM	2,00	15	29,8	39,2	44,2	[illegible]	51,2	58,1	21,9	27,2	29,5	31,7	42,1	48,1	41	139	293
			733	964	1087	[illegible]	1260	1429	377	468	508	546	725	828			
BE 300 ERMR below	1,50	7,5	23,0	30,1	31,8	[illegible]	39,1	42,7	17,1	20,9	22,4	24,8	31,0	33,9	32	90	178
			566	740	782	[illegible]	962	1050	294	360	386	427	534	584			
BE 300 ERMR above	1,00	1,8	16,6	20,2	21,8	[illegible]	26,7	29,1	12,2	14,4	15,7	18,1	21,7	23,6	20	58	121
			408	497	536	[illegible]	657	716	210	248	270	312	374	406			
BE 400 ERMR below	1,80	11	27,2	34,8	38,9	[illegible]	45,1	50,7	20,4	25,5	27,5	29,3	37,1	41,2	53	114	210
			669	856	957	[illegible]	1109	1247	351	439	474	505	639	709			
BE 400 ERMR above	1,00	3	16,7	20,0	21,5	[illegible]	26,1	28,2	12,4	14,5	15,4	18,0	21,4	23,0	12	40	83
			411	492	529	[illegible]	642	694	214	250	265	310	369	396			
BE 500 ERMR below	2,00	15	29,8	39,2	44,2	[illegible]	51,2	58,1	21,9	27,2	29,5	31,7	42,1	48,1	41	139	293
			733	964	1087	[illegible]	1260	1429	377	468	508	546	725	823			
BE 500 ERMR above	1,00	3,7	16,2	19,6	20,9	[illegible]	25,0	27,5	11,4	13,5	14,0	16,8	19,9	21,0	19	55	109
			399	482	514	[illegible]	615	677	196	232	241	289	343	362			

As the piping is usually short, an additional 50 per cent or so should be added to cover the moulded parts.

The pressure loss over the heat exchanger(s) in the hot water tank is listed by the manufacturer. Where this is not listed, then the manufacturer should be consulted, or the loss calculated using the data that is provided where content and diameter of the heat exchanger(s) is known.

TIP

Bivalent hot water tanks are often installed for hot water generation, their heat exchange coils connected in series. Where this is the case, both pressure losses must be added.

Here, we take a bivalent 300-litre tank as an example.

When we know the individual pressure losses, we can calculate the total pressure loss ΔP_H using the equation above.

TIP

If the hot water tank and buffer tank each have a separate charging pump, then there is no pressure loss over the switching valve. In this case, we need to consider pressure loss over the non-return valve, although this will generally be less than over the switching valve. This has the additional advantage that, during heating operations, we only need to consider the almost inconsequential pressure loss over the buffer tank. This saves both energy and costs.

7.2.1 Example: Dimensioning the secondary pump or charging pump

Let us stay with our example of a brine-water heat pump with an output of 12.4 kW. According to the manufacturer, we need a minimum flow rate of 2.6 m³/h. To calculate total pressure loss:

$$\Delta P_H = P_{HPC} + P_{PH} + P_{SV} + P_{MBH} + P_T$$

The pressure loss across the condenser for the selected heat pump is:

$$P_{HPC} = 0.20 \text{ bar}$$

We lay a total of 12 m of 1″ piping (corresponds to PE DN = 32). If we refer to the table for pressure loss for PE pipes (Table 7.1), we can calculate the pressure loss based on the width and the flow rate:

$$P_{PH} = 8 \text{ mWs/100m} \times 12 \text{ m} = 1.32 \text{ mWs} = 0.1$$

The pressure loss for the three-way switching valve can be determined using the K_{VS} value and the corresponding diagram. According to the manufacturer's specifications, the 1″ three-way switching valve has a K_{VS} value of 5.7.

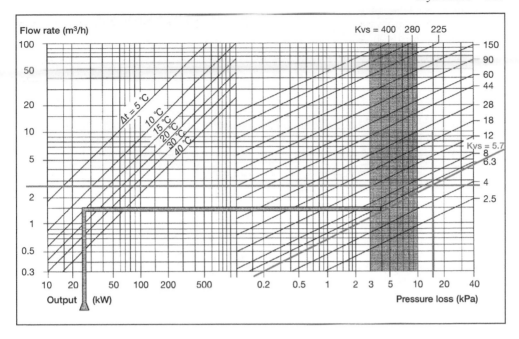

Figure 7.9 Determining pressure loss across a three-way switching valve

According to the diagram, the pressure loss is around 8 kPa.

$$1 \text{ bar} = 10^5 \text{ Pa} = 10^2 \text{ kPa}$$

Thus:

$$8 \text{ kPa} = 0.08 \text{ bar}$$

To calculate pressure loss across moulded parts we assume a maximum of six moulded parts.
    This corresponds to a total length of 1.2 m × 6 = 7.2 m. Therefore, the pressure loss across six moulded parts is:

$$P_{MBH} = 5 \text{ mWs}/100 \text{ m} \times 7.2 \text{ m} = 0.36 \text{ mWs} = 0.04 \text{ bar}$$

Using Table 7.3, we take the pressure loss across the tank as 2 m³/h and a bivalent hot water tank with 300-litre capacity:

$$P_T = 90 \text{ mbar} + 58 \text{ mbar} = 148 \text{ mbar} = 0.15 \text{ bar}$$

Therefore, for the total pressure loss:

$$\Delta P_H = P_{HPC} + P_{PH} + P_{SV} + P_{MBH} + P_T$$
$$= 0.20 \text{ bar} + 0.06 \text{ bar} + 0.08 \text{ bar} + 0.04 \text{ bar} + 0.15 \text{ bar}$$
$$= 0.53 \text{ bar}$$

With the minimum flow rate (1.9 m³/h) and pressure loss, we can now select the best charging pump for the heating side.

These values can now be used to select the brine circulation pump:

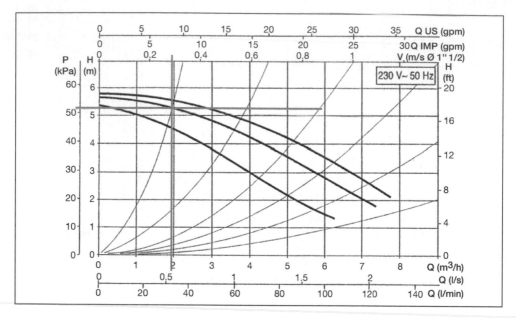

Figure 7.10 Determining the heating circulation pump

TIP

This heating circulation pump must be set at level 2.

This is for charging pump model A 50.180 M.

Now let us look at the same heat pump system, this time without a switching valve. Instead, we have two charging pumps – one for the hot water tank, and one for the buffer tank.

The pressure loss across the non-return valve is roughly twice as much as across a moulded part.

Therefore, the pressure loss across the non-return valve is calculated across the equivalent length of:

$$L = 2 \times 1.2 \text{ m} = 2.4 \text{ m}$$

As: $P_{NRV} = 5 \text{ mWs}/100 \text{ m} \times 2.4 \text{ m} = 0.12 \text{ mWs} = 0.01 \text{ bar}$

Therefore, for the charging pump on the hot water tank:

$$\Delta P_H = P_{HPC} + P_{PH} + P_{NRV} + P_{MBH} + P_T$$

$$= 0.20 \text{ bar} + 0.06 \text{ bar} + 0.01 \text{ bar} + 0.04 \text{ bar} + 0.15 \text{ bar}$$

$$= 0.46 \text{ bar}$$

Therefore, for the charging pump for the hot water tank:

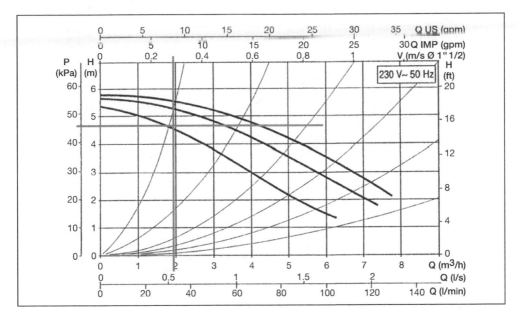

Figure 7.11 Determining the heating circulation pump for the hot water tank

---

TIP

This charging pump can be set at level 1. This saves electricity and costs for hot water generation.

This applies to charging pump model A 50/180 M.

For the charging pump for the buffer tank, P_{BT} = 0 bar:

$$\Delta P_H = P_{HPC} + P_{PH} + P_{NRV} + P_{MBH} + P_{BT}$$

$$= 0.20 \text{ bar} + 0.06 \text{ bar} + 0.01 \text{ bar} + 0.04 \text{ bar} + 0 \text{ bar}$$

$$= 0.31 \text{ bar}$$

Therefore, for the buffer tank charging pump, see Figure 7.12.

TIP

This heating charging pump should be set at level 3, saving even more electricity and costs for loading the buffer tank.

This applies to the charging pump model VA 65/130-1/2".

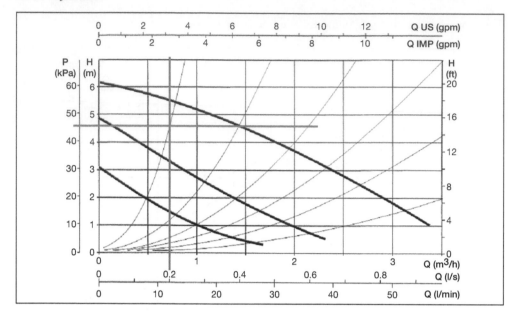

Figure 7.12 Determining the heating circulation pump for the buffer tank

The hydraulics for this heat pump system are as follows:

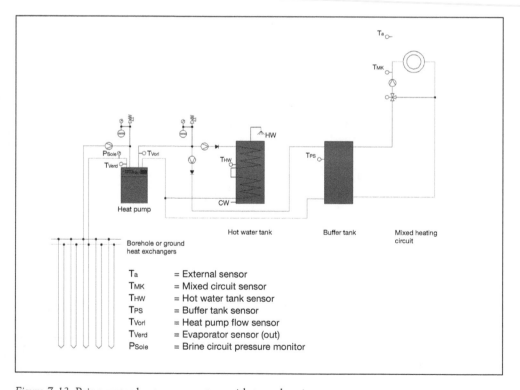

Figure 7.13 Brine-water heat pump system with two charging pumps

The values calculated above show very clearly that this heat pump system is even more energy efficient because the pressure loss the pumps must overcome is lower. This is particularly the case for heating operations, the main operating factor in a heat pump. A lower pressure loss means less physical work and therefore lower energy expenditure.

Energy efficient pumps that consume even less energy than standard circulation pumps are available. These energy efficient pumps should be used wherever long running times are expected. This is less often the case for charging pumps. However, as the heating circuit pumps do have long operating times, it is worth using these efficient pumps.

7.3 The buffer tank

A heat pump should always be operated using a buffer tank! Why?

A buffer tank stores/buffers heat and simultaneously functions as a hydraulic separation between the heat pump and the rest of the heating system. A buffer tank is essential to ensure the flow volume needed by the heat pump. The volume flow rate of a heat pump is usually significantly greater than the volume flow rate on the heating side.

The buffer tank also extends the lifespan of a heat pump by making long running times and long breaks in operation possible, especially during the transition periods in spring and autumn. During these periods, the heat pump supplies much more energy than is needed for heating.

A buffer tank also increases the 'heating comfort' by enabling a better temperature balance to be established within the building (e.g. when the sun shines in or when individual heating circuits are turned down by individual room thermostats). This allows the room temperature of individual rooms to be set and maintained at specific temperatures. This is not possible for all rooms if there is no buffer tank.

It is also possible to create 'cheaper' heat pump systems by using the screed flooring as a buffer, rather than having a buffer tank. A heat pump needs a storage medium to ensure a so-called minimum operating period. If the floor screed is to serve as the storage medium, then the individual room thermostats for underfloor heating, which are otherwise mandatory, must be at least partly abandoned. In this case, the heat pump transfers heat

Figure 7.14 Buffer tank

to the open underfloor heating circuit. However, this involves a loss of comfort as it is no longer possible to directly control room temperature.

TIP

German law covering heating systems makes individual room regulation mandatory.

However, there is (unfortunately) currently an exception for heat pump systems. But individual room control is state of the art, and therefore should be used.

When sizing the buffer tank, we can take 20–30 l/kW as a guide value. A 300 l buffer tank is recommended for a heat pump with an output of 12 kW. It is generally not a mistake to select a somewhat larger buffer tank.

Many homebuilders wish to install a fireplace with a stove, and this makes ecological sense. These stoves quickly emit heat of 8 kW or more. Thanks to the German Energy Saving Ordinance (ENeV), all newly built houses in Germany are well insulated. Consequently, a stove can quickly cause a building to overheat. Therefore, manufacturers offer stoves with heat exchangers, so that much of the heat can be transferred to a buffer tank, and then used as required.

However, the recommended guide values no longer apply when sizing the buffer tank for a stove. The buffer tank must be dimensioned in accordance with the stove output. How the stoves are used is also important: if run more frequently and at hotter temperatures, then the size of the buffer tank increases. A buffer tank with a volume of 1,000–1,500 l is recommended for detached houses.

TIP

A mixed heating circuit behind the buffer tank is absolutely essential for surface heating. This allows the heating flow temperature to be adjusted according to the weather conditions, and avoids overheating.

Figure 7.15 shows how a stove can be integrated into a heat pump system with free cooling.

Depending upon heat output and customer requirements, the stove can also be used for hot water generation: this makes economic sense because the stove would always provide high flow temperatures. However, this does require greater expense when it comes to controller technology, including an additional three-way switching valve and a regulator.

TIP

The controller for the stove should be sourced from the stove manufacturer, because the exhaust gases are subject to a minimum permitted temperature.

The controllers should have permanent temperature limits to avoid overheating. A mixed heating circuit is also necessary to regulate the flow temperature.

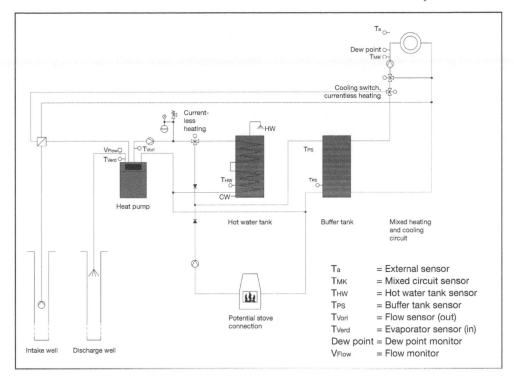

Figure 7.15 Water-water heat pump system with stove connection + cooling

A buffer tank offers the following key advantages:

1   Hydraulic separation between heat pump and heating circuit.
2   The heat pump turns on and off less frequently (i.e. it has longer operating and standing periods), which significantly extends its lifespan.

Figure 7.16 Buffer tank

3   Greater comfort, because the temperature can be set individually in each room. Free cooling is also possible, using a cooling ceiling controller for heating and cooling.
4   The heat can be released by the buffer tank as needed, offering maximum comfort in the rooms being heated.
5   If the buffer tank is sufficiently large, a stove can be connected to the heating system.

However, point 5 can only be fulfilled when the buffer tank is connected as shown in Figure 7.15. The flows through the heat pump and through the heating circuit can certainly vary, especially during transitional periods. Yet one often sees buffer tanks connected as follows:

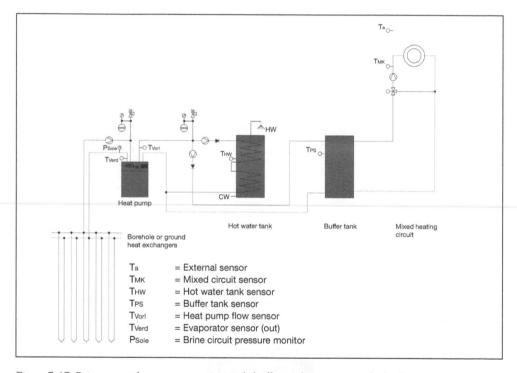

T_a	= External sensor
T_{MK}	= Mixed circuit sensor
T_{HW}	= Hot water tank sensor
T_{PS}	= Buffer tank sensor
T_{Vorl}	= Heat pump flow sensor
T_{Verd}	= Evaporator sensor (out)
P_{Sole}	= Brine circuit pressure monitor

Figure 7.17 Brine-water heat pump system with buffer tank in series with the heating circuit

When the buffer tank is connected with the heating circuit in series, point 1 no longer applies: hydraulic separation is impossible because the flow through the heat pump and through the heating circuit is one and the same.

However, only one pump is required if connected using this arrangement. The pump serves both as a charging and a circulation pump.

Consequently, the charging pump (which is simultaneously the circulation pump) must run constantly, ensuring supplies to each heating circuit and hot water generation. The pressure losses are greater and, because the pump is continually in operation, power consumption is also higher.

When connected as shown above, rooms are not supplied with heat when hot water is being generated. For well-insulated new builds, however, this is not a problem.

To summarise, this form of connection saves on one pump, but does involve a noticeable increase in power consumption.

7.4 Hot water generation

There are several methods of hot water generation:

1   hot water generation using an instantaneous water heater;
2   hot water generation using an electrical heating element in a separate hot water tank;
3   hot water generation using the heat pump via a hot water tank;
4   hot water generation with a heat pump and an additional solar system; and
5   hygienic hot water generation based on the continuous flow principle.

Each of these options has its advantages and disadvantages, and these must be considered. We look at them all below. Generally, if a heat pump is installed, then generating hot water with an instantaneous water heater or electric heating element usually makes *no sense*. Direct electrical hot water generation is around three times more expensive than using a heat pump!

7.4.1 Hot water generation using an instantaneous water heater

For the reasons given above, generating hot water using an instantaneous water heater makes no sense, particularly for smaller housing units.

Generating hot water using a decentralised instantaneous water heater can make sense, or indeed be necessary, for larger buildings such as apartment blocks with multiple housing units. Special high-performance hot water tanks with large hot water draw-off capacities (German standard DIN 4708 performance index N sub L – 'NL number') are available for generating hot water using larger heat pumps for use in apartment blocks. See Section 7.4.3, 'Hot water generation using the heat pump via a hot water tank'.

IMPORTANT

Note that for storage volumes over 400 l, the entire volume of hot water must be regularly heated to above 60 °C as a means of protecting against legionnaires' disease. As this is only possible using an additional electric heating element, it makes it significantly less economical to generate these large quantities of hot water with a heat pump.

Centralised hot water generation using a hot water tank requires piping for hot water and cold water. This increases the hot water tank installation costs accordingly.

If the demand for hot water is such that additional electrical heating to 60 °C is required once a day, then it is worth considering using an instantaneous water heater to generate hot water. This saves the cost of a hot water tank (plus space savings) and simplifies pipe installation as hot water pipes, circulation pipes and insulation are no longer required. In larger buildings, this also avoids the heat loss incurred through hot water circulation and piping, which can be considerable. A further advantage is the availability of decentralised, instantaneous, hygienic hot water generation – with no hot water tank.

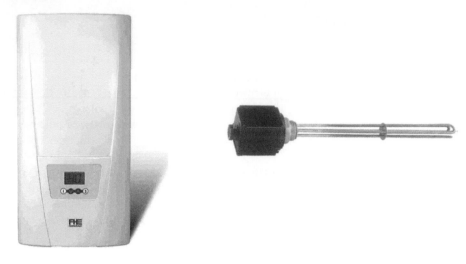

Figure 7.18 Instantaneous water heater   *Figure 7.19* Electric heating element

7.4.2 Hot water generation using an electric heating element in a separate hot water tank

This form of hot water generation makes no ecological or economic sense. If the water is heated within a storage tank, then this should be done by the heat pump because the costs of generating hot water with an electric heating element are at least three times higher.

- *This form of hot water generation is only advantageous for the seller of cheaper equipment who can offer their own product more cheaply.*
- *In a heat pump system, it should not be possible to switch on an electrical heating element via a controller!*

If an electrical heating element is installed, then it should only be possible to turn it on by hand when and if it is needed. If the electrical heating element is turned on by the controller, then this removes any operator control. This can lead to nasty surprises when the electricity bill arrives!

7.4.3 Hot water generation using the heat pump via a hot water tank

This is the traditional and most sensible method of generating domestic hot water for smaller housing units. The heat pump provides sufficient heat for heating the house and for generating domestic hot water in an extremely efficient manner.

For smaller detached houses, a 300 l or 400 l bivalent hot water tank is usually sufficient. Smaller monovalent hot water tanks are not suitable; these are usually only equipped with a single pipe heat exchanger, which is far too small for transferring the quantities of heat involved. If the heat pump cannot transfer all the heat it produces, this can lead to a high pressure fault and the heat pump being shut down.

For larger heat pumps, we need to consider the relatively high nominal flow rate on the heating side, the associated pressure loss and the transfer of the heat output. Here, heat

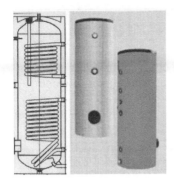

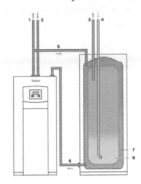

Figure 7.20
Bivalent hot water tank

Figure 7.21
High-performance hot water
tank

Figure 7.22
Double-walled hot water tank

pumps behave in a completely different way to instantaneous water heaters; as heat pumps
only work with low flow temperatures, a much larger exchanger surface area is required to
transfer the heat they generate. Therefore, bivalent hot water tanks should only be used
for smaller heat pump systems up to around 12 kW.

The flow resistance across the tank and the heat transfer via a correspondingly large
heat exchanger are especially important in larger heat pump systems (e.g. 12 kW and above,
depending on manufacturer): these higher flow resistances lead to correspondingly higher
energy losses across the charging pump, and an undersized heat exchanger prevents sufficient
heat transfer away from the heat pump. This is why hot water tanks with very low flow
resistances and heat exchangers with particularly large surface areas (see Figure 7.20) have
been developed for heat pump systems. The hot water tank shown on the left has a pipe
cluster double heat exchanger with a large surface area, and the one shown on the right
is a double-walled hot water tank.

IMPORTANT
- It is important to have a sufficiently large heat exchanger in the hot water tank!

If the heat exchanger is too small, then the heat pump cannot transfer enough heat. This
results in a high pressure fault and the shutting down of the heat pump. A bivalent hot
water tank is sufficient for smaller heat pumps (e.g. for a detached house) (< 12 kW). Larger
heat pump systems require high performance hot water tanks with especially large heat
exchangers and low flow resistances that are able to absorb a correspondingly large heat
output.

- *A sufficiently large hot water tank is also important!*

The hot water tank must be large enough because, compared to a standard instantaneous
water heater, the heat pump heats using significantly lower flow temperatures. This is
compensated by using a larger heat exchanger and a greater volume of hot water.

TIP

Servicing the hot water tank

As a heat pump operates with no need for servicing, all the other components should also be as maintenance-free as possible. Therefore, a sacrificial anode is recommended.

The sacrificial anode provides continual and maintenance-free corrosion protection.

Figure 7.23 Sacrificial anode

For larger heat pump systems where the hot water flow on the heating side must be greater than 4–5 m³/h, hydraulic separation via a heat exchanger is recommended:

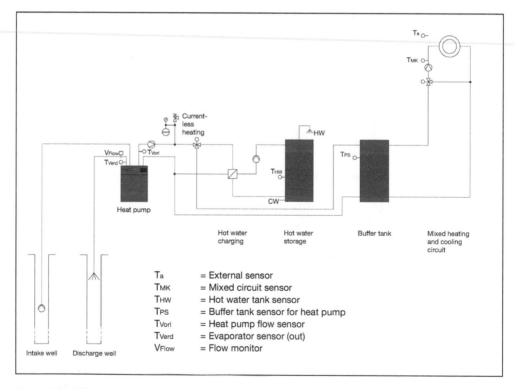

Figure 7.24 Water-water heat pump system with storage tank charging system

A storage tank charging system allows the required and optimal nominal flows for the heat pump and the charging storage tank to be synchronised.

7.4.4 *Hot water generation with a heat pump and an additional solar system*

Normally a bivalent hot water tank is used for solar hot water generation. The solar system is connected to the lower heat exchanger, and the oil or gas-fired boiler to the upper heat exchanger. This arrangement cannot always be transferred to a heat pump system and can only be achieved with smaller heat pumps.

• *The heat pump usually needs both heat exchangers.*

This leads to two legitimate questions:

1 Can a solar system be used in combination with a heat pump?
2 Is that not a bit too much of a good thing?

The answers:
To the first: solar power? Of course! A heat pump system can also use solar power.
The answer to the second question: no!

The heat pump has to work with a higher flow temperature to generate domestic hot water. Therefore, the refrigerant in the heat pump must be highly compressed to achieve the high temperatures. (This is also the reason that the performance level – COP – is lower during hot water generation.) This in turn leads to higher power consumption and thus higher energy costs. Therefore, combining solar-powered hot water generation with a heat pump makes sense. Electricity costs will continue to rise but we certainly are not going to get electricity bills from the sun!

Figure 7.25 Bivalent solar storage tank, designed for heat pumps

So it is possible to use solar power to support the heating system. Yet customers do need to be aware of the long amortisation periods involved.

There are different ways of using solar power, and we demonstrate two of them below.

Bivalent solar storage tanks designed especially for use with heat pumps were shown at the 2009 ISH trade fair for water and energy in Frankfurt.

Just like the high-performance hot water tanks, these bivalent tanks contain a double, smooth pipe heat exchanger with a large surface area (not visible in Figure 7.25) with similar characteristics to the high-performance hot water tanks described above. These tanks are also available in a variety of sizes.

Another option is to connect the solar system to an external plate heat exchanger and use the low-flow technique. A high-performance hot water tank makes greater draw-off capacities possible.

The following diagram shows a heat pump system with solar-powered hot water generation and a special bivalent solar tank for heat pumps:

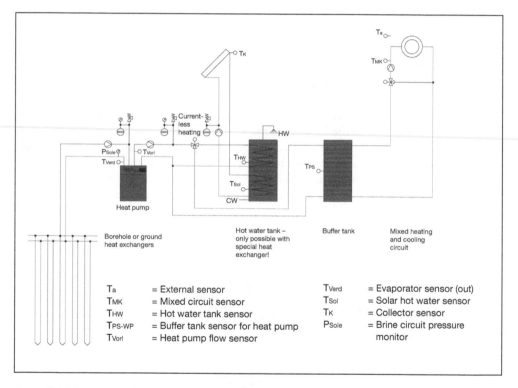

Figure 7.26 Brine-water heat pump system with solar hot water generation

This heat pump system generates hot water in the normal manner via the heat pump, and uses the extra solar power as an auxiliary heat source.

PLEASE NOTE

Do not forget the anti-scald mixing valve when using solar power systems!

7.4.5 *Hygienic hot water generation using the continuous flow principle*

A heat pump system can also be used for hygienic hot water generation, although this involves complex technical measures. The operating characteristics of a heat pump prevent optimal layering in a central buffer tank, so reducing the efficiency of the heat pump system as a whole. Optimal operation requires two separate buffer tanks + hygienic hot water generation. This combination is rarely used for reasons of expense.

7.5 Dimensioning the hot water piping

It is important that the hot water piping for heat pumps is correctly sized. The hot water temperature is usually lower than with instantaneous water heater systems, and so it is recommended that pipes of a greater diameter are used to compensate for this. This ensures that roughly the same quantity of heat is available at the tap as there is with an instantaneous water heater.

Mixing faucets regulated by thermostats are recommended for additional comfort in the bathroom.

7.6 Electronics and heat pumps

A heat pump system also needs the correct electrical connections to work properly. The electrical connections are arranged across several circuits: we start with the power supply, because a heat pump cannot operate without electricity. This is usually divided into 400 V AC (power current) for the pumps and the compressor, and 230 V AC for the control voltage. There are also individual control cables for smaller pumps, switching valves, probes and sensors.

7.6.1 Power supply

The manufacturer's specifications must be adhered to when determining the power supply, especially the compressor starting current, which is a multiple of the nominal current.

Overall, the following points must be noted:

1. Fuses:
The main power supply, usually 400 V AC, must be protected using time-lag fuses. Fuses that react too quickly will blow each time the compressor starts, making it impossible to operate the heat pump. Note the manufacturer's specifications.
Three-phase power supply requires fuses that disconnect all the poles when blown.

The control voltage is usually supplied in a single phase and separately protected with its own circuit breaker.

2. Soft starter:
On larger heat pumps, it can be a good idea to turn on the compressor (or compressors) using a soft starter, especially in rural areas where power is supplied via long overhead power lines.

During the planning phase, the power supply conditions should be checked with the local utility.

3. Feeding in 400 V AC electric current and control voltage:
Many utilities demand that all poles be disconnected during the supply blocking times. Once the blocking period is over, the utility sends a control signal indicating that power is again available, and a power circuit breaker (fuse) simultaneously turns on the power voltage and sends the controller the signal for the heat pump to start operating again. This allows the controller to safely restart the heat pump.

7.6.2 Control voltages

The controller delivers control voltages, usually 230 V AC, for actuators (smaller pumps and switching valves). These actuators must be connected according to the manufacturer's terminal and circuit diagrams. It is extremely important that the maximum permitted loads (i.e. currents) are not exceeded. Overloading makes the contacts on the relays wear faster, which in turn can lead to faults in the outlets. When this happens, the controller or relevant module must be exchanged. Where there is inductance (i.e. for electric motors (pumps)), it is important that the maximum permitted currents are not exceeded. Inductance exerts significantly more wear on the contacts than purely resistive loads.

7.6.3 Monitors and sensors

Sensors such as flow and pressure monitors and temperature sensors are needed for regulating and controlling the heat pump. These sensors supply the controller with the current values. The controller then compares these values against programme settings and ideal values before opening or closing the relevant control elements, and turning on or off actuators such as pumps and control elements, as required.

8 Heat pump system planning guidelines

The following table offers guidelines for planning a heat pump system:

Table 8.1 Planning sheet for heat pump systems

Building type				
Existing stock	Flow temperature:	Higher than 50 °C (55 °C)	Heat pump generally not recommended	
		Lower than 50 °C	Only underfloor heating	Heating load calculated by energy consultant
			Underfloor heating and radiators	
New build	Heating output	$P_H = $ ___ m² × 45 W/m² = _____ W		$P_H = $ _____ kW
	Output for hot water generation	$P_{HW} = (0.25 - 0.35)$ kW/People × ___ People = __ kW	$P_{HW} = $ _____ kW	
	Total heat output	$P_{H'} = P_H + P_{HW} = $ ___ kW + ___ kW = ___ kW		$P_{H'} = $ _____ kW
	Factor for utility blocking time	$P_{H''} = $ ___ × $P_{H'} = $ ___ × ___ kW		$P_{H''} = $ _____ kW
Type of heat pump				
Water-water heat pump	Water supply sufficient?	No		
		Yes ──		
	Discharge capacity sufficient?	No	No water-water heat pump	
		Yes ──		
	Water quality OK?	No		
		Yes ──→ Water-water heat pump possible		
	Selecting the charge pump	Minimum flow rate through the evaporator according to manufacturers specifications:	___ m³/h	
		Pressure drop over pumped height:	___ mWs = ___ bar	
		Pressure loss over feed PE pipes:	___ mWs = ___ bar	
		Pressure loss over return PE pipes:	___ mWs = ___ bar	
		Pressure loss over moulded parts (bends, etc.):	___ mWs = ___ bar	

Table 8.1 Planning sheet for heat pump systems—*continued*

Water-water heat pump	Selecting the charge pump	Pressure loss over the evaporator according to manufacturer specifications:		___ kPa = ___ bar
		No free cooling		
		With free cooling:	Pressure loss over heat exchanger with cooling:	___ kPa = ___ bar
		Total pressure loss:		___ bar
		Choice of suitable charging pump:		
Brine-water heat pump	Ground heat exchangers	Sufficient space?	No	No ground heat exchangers
			Yes	
		With free cooling?	Yes	
			No	Worth using ground heat exchangers
	Borehole heat exchangers	Sufficient space?	No	No borehole heat exchangers
			Yes	Borehole heat exchangers possible
	Selecting the brine circulation pump	Minimum flow rate through the evaporator according to manufacturer specifications:		___ m³/h
		Pressure loss over borehole heat exchangers:		___ mWs = ___ bar
		Pressure loss over brine circuit manifold:		___ mWs = ___ bar
		Pressure loss over feed PE pipes:		___ mWs = ___ bar
		Pressure loss over return PE pipes:		___ mWs = ___ bar
		Pressure loss over moulded parts (bends, etc.):		___ mWs = ___ bar
		Pressure loss over the evaporator according to manufacturer specifications:		___ kPa = ___ bar
		No free cooling		
		With free cooling:	Pressure loss over heat exchanger with cooling:	___ kPa = ___ bar
		Total pressure loss:		___ bar

Table 8.1 Planning sheet for heat pump systems—*continued*

		Choice of suitable brine circulation pump:	
Air-water heat pump	Compact unit		
	Split unit	Consider level of noise emissions	
Selecting the buffer tank		Heat pump output P_H: ___ kW	
		Volume V = 25 l/kW × P_H = 25 l/kW × ___ kW =	___ l
Selecting the hot water tank		1–2 detached houses for up to six people with normal consumption patterns:	300 l
		1–2 detached houses for up to eight people with normal consumption patterns:	400 l
		Apartment blocks:	High-performance hot water tank 300 l
			High-performance hot water tank 400 l
			Storage tank charging system
			Instantaneous water heater
Choosing the charging pump(s)		Minimum flow rate through the condenser according to manufacturer specifications:	___ m³/h
		Pressure loss over the condenser according to manufacturer specifications:	___ kPa = ___ bar
		Charging pump + switching valve: / Pressure loss over switching valve:	___ kPa = ___ bar
		Pressure loss over piping:	___ mWs = ___ bar
		Pressure loss over hot water tank:	___ mbar = ___ bar
		Total pressure loss:	___ bar
		Choice of suitable heating pump:	
		Charging pump for hot water tank: / Pressure loss over piping:	___ mWs = ___ bar
		Pressure loss over hot water tank:	___ mbar = ___ bar
		Total pressure loss:	___ bar
		Buffer tank charging pump = pressure loss over piping:	___ mWs = ___ bar

9 Different types of heat pump system

Heat pump systems should be chosen for reasons other than simply the heat output needed by the house and its inhabitants.

- The entire heat pump heating system must reflect the wishes and needs of the operator, as for all other house installations.

Homeowners usually wish to heat a building as economically as possible. To achieve this, you need answers to the following questions:

- What type of heat pump does the customer want? A water-water heat pump, or a brine-water heat pump with borehole or ground heat exchangers? Or an air-water heat pump?
- Will the house have underfloor heating or radiators?
- Does the owner want a heat pump system to be connected to a stove (often clearly discernible on the building plan)? If so, then a larger buffer tank is recommended!
- Does the owner also want the heat pump to cool the house? This will make it much more comfortable to live in.
- Does the owner want to use generate domestic hot water using solar power? This requires a larger hot water tank and the low-flow technique.
- Should the heat pump be connected to a swimming pool? A heat pump is ideal for heating a swimming pool as the flow temperatures are usually very low, allowing the heat pump to work at maximum efficiency.
- Are extensions to the building planned for a later date?
- Does the garage also need to be heated?

Some customers have very definite ideas while others are looking for specialist advice. Here, we first need to consider what is possible and what is sensible according to the particular circumstances.

A heat pump can only work at its best when the house plans include underfloor heating using low possible flow temperatures. If the homeowner is planning to use radiators, then he should be informed of the advantages of underfloor heating or wall surface heating.

If the house plans include a stove, then homeowners are advised to choose a stove with a water connection. This will then need to be connected to a correspondingly large buffer tank. This is the only way in which a large part of the heat generated by the stove can be transferred to the buffer tank and then used as required in rooms requiring heating. Stoves without water connections are unsuitable in increasingly well-insulated houses because the rooms quickly become overheated.

Cooling in summer is a welcome by-product of a heat pump. This makes the climate within the house more pleasant and at minimal extra cost. Naturally, this is a decision for the individual homeowner.

Solar-powered hot water generation makes sense because the efficiency of a heat pump falls as the flow temperature rises. Consequently, the electricity consumed for generating domestic hot water is greater than for heating, although it is still much less expensive than using an electric heating element. Therefore, solar-powered hot water generation is generally a good idea.

If the customer plans to connect the heat pump to a swimming pool, then the pool should also be heated by a heat pump. The relatively low water temperatures in a swimming pool allow the heat pump to work at maximum efficiency.

If the housebuilder is planning to extend the house at a later stage (attic conversion, other additions, etc.), then it is worth taking this into account when planning the heat pump. The probability of later extensions should also be considered, although you cannot consider every eventuality.

Heating a garage can be a sensible investment. It reduces humidity and therefore corrosion, especially in wet and cold weather. In Germany, garage heating is subject to separate regulations, and a garage may not be heated in the same manner as rooms in a house.

9.1 A simple water-water heat pump system

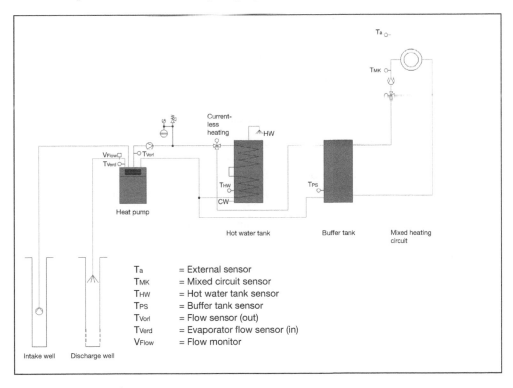

Figure 9.1 Water-water heat pump system

This simple water-water heat pump system is used to heat a residential building and generate hot water using a bivalent hot water tank and a buffer tank with mixed heating circuit. The individual rooms can each be equipped with their own thermostats.

9.2 A brine-water heat pump system with free cooling

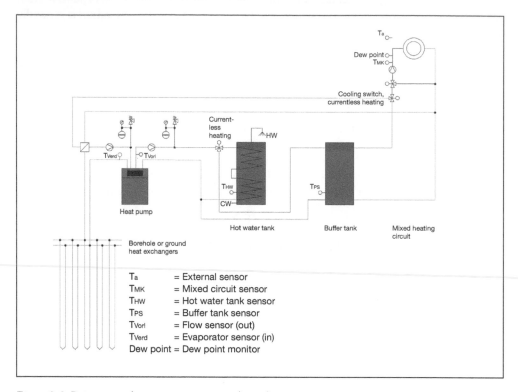

Figure 9.2 Brine-water heat pump system with cooling

Using this heat pump system, the operator can cool a house at no great extra cost. As free cooling is possible at such low prices, heat pump systems with cooling are becoming more frequently installed.

The cooling system can be optimised by using two separate circuits: a heating circuit and a separate cooling circuit with underfloor heating and ceiling cooling. Heat rises and cold falls.

By using cooling ceiling controllers, rooms can be heated or cooled on an individual basis.

9.3 A brine-water heat pump system with free cooling, solar system and closed stove

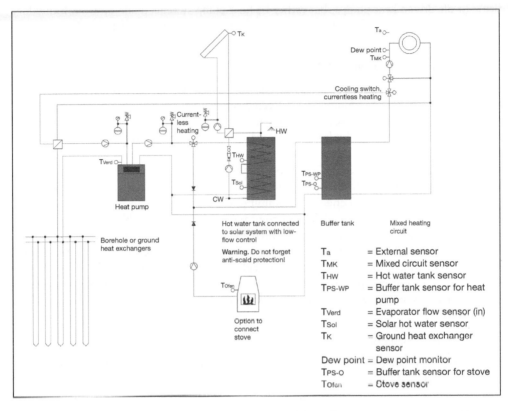

Figure 9.3 A brine-water heat pump system with free cooling, solar-powered hot water generation and stove connection

This heat pump system meets the most demanding of requirements. It can provide cooling on hot summer days, and heating on cold days. Connecting it to a solar system and a stove significantly increases the overall efficiency (COP) of the heating system. On sunny days in summer, the heat pump often remains off, or is only seldom turned on. When the stove is lit, a large part of the heat it produces is transferred to the buffer tank and then to individual rooms, as required. This requires mixed heating circuits in order to compensate for fluctuating temperatures.

Here, the solar system operates according to the low-flow principle. A solar system can also be directly connected using a special bivalent solar storage tank for heat pumps – hot water generation with a heat pump and an additional solar system.

This heat pump system can be extended as required (e.g. with a further heating circuit, heating for a swimming pool or garage, etc.).

9.4 Types of heat pump

Several types of heat pump are available. Here, we do not differentiate between water, brine or air-driven heat pumps, but rather by heat pump systems made up from individual components, one of which is the actual heat pump, and compact systems.

9.4.1 Stand-alone heat pump units

These are heat pumps consisting only of the actual heat pump. The hot water tank and buffer tank, as well as the pumps, are delivered separately.

Heat pump systems composed of individual components are more work to install because each component must be hydraulically and electrically integrated. Additionally, the components must be carefully checked to determine their compatibility, and having individual components takes up more space. However, the overriding advantage of this heat pump system is the clear overview it provides, making replacement of any defective components both easier and cheaper. Obtaining replacement parts is also simple – even if the original part is no longer available, it can be replaced by a compatible equivalent. Where a fault does occur, this can often be repaired by a general HVAC technician. If a tank becomes defective (after several years this usually happens to the hot water tank – even stainless steel tanks), then it can simply be replaced with a new one.

Examples of heat pump systems consisting of a stand-alone heat pump plus hot water and buffer tanks are shown in Chapter 17, 'Examples of heat pump systems'.

9.4.2 Compact systems

Compact systems are compact heat pump units consisting of the actual heat pump plus a hot water tank and the relevant pumps.

Compact heat pumps are easy to plan and install because the heat pump system (consisting of the heat pump unit, hot water tank and usually the pumps) are all factory-made and ready for installation.

The disadvantage is its dimensions: all the components are contained within the heat pump housing, and are therefore as small as possible. As a result, there is usually no room for a buffer tank. This is a disadvantage because the heat pump is forced to stop and start more frequently, resulting in greater wear and tear. Furthermore, as technology changes so rapidly, obtaining spare parts in the future may not be so simple. The brine and charging pumps are usually integrated into the compact systems and so care must be taken that the borehole or ground heat exchangers are dimensioned accordingly and the pressure losses correctly calculated.

Usually, compact systems include electric heating elements to support the heat pump as required. This often leads to large electricity bills for the operator.

The advantage is, of course, that compact heat pump systems require less space and involve less installation work.

The disadvantage is that after several years in operation, replacement parts for components within the system may no longer be available.

We are familiar with this phenomenon from the compact music systems from the 1970s and 1980s. When a single part became defective, often there was no alternative but to scrap the entire system. I can remember my father buying such a system. He was terribly proud of it, until the record player no longer worked and no replacement parts were available. The whole unit got thrown out. Since then, he has only bought stand-alone devices.

It has now become standard practice in the entertainment industry to build stand-alone devices. Has this taught us nothing when it comes to heat pumps? Obviously not, because complete heat pump systems are being offered as compact units. But what happens after 10 years, for example, when the hot water tank, which is often positioned directly above the heat pump components, corrodes and leaks? The entire system is destroyed. Repairs on these systems are also much more difficult and – where possible at all – more expensive. They may also involve the expense of calling out the manufacturer's own technicians. And yet the repair again depends upon replacement parts being available.

Heat pump systems built from individual components, such as the heat pump, a separate hot water tank and a buffer tank, may have the disadvantage of greater installation effort and therefore costs, but the advantage is obvious: any faulty component can be exchanged at any time. Replacement parts can often be purchased in standard outlets. This also cuts maintenance costs substantially.

Figure 9.4 A compact system

The following table indicates the advantages and disadvantages of heat pump systems consisting of individual components and compact heat pump systems:

Table 9.1 Comparison between individual components and compact systems

Heat pump systems with individual components		Compact systems	
++	Easy to get replacement parts	–	Availability of replacement parts uncertain, especially hot water tank
+	Larger hot water tank possible, guideline: 300 l	–	Very small hot water tank
+	Usually with buffer tank	–	Often without buffer tank
–	More complicated installation	+	Simple to install
–	Requires more space	+	Needs little space
+	Good overview of system components	–	Difficult for non-specialists to service/repair
+	Hydraulics can be better adjusted	–	Subsequent parts must be hydraulically adapted to the integrated pumps
–	Heat pump can be somewhat louder	+	Compact heat pump systems can be quieter

10 Economic considerations

10.1 Is a heat pump worthwhile economically?

This is a frequently asked question, and one that can be answered simply:

Yes, of course!

Sadly, Germany is currently a less developed country when it comes to heat pumps; there must be a reason that 90 per cent of all new builds in Switzerland are equipped with heat pumps!

For new builds especially, there is no meaningful alternative to a heat pump (assuming the underfloor heating is adequately dimensioned, of course). Since Germany's renewable energy act (EEWärmeG) came into force, a heat pump system is only marginally more expensive than a combined heating system based on a gas boiler and solar.

Heat pumps are suitable in existing buildings if there is a low temperature heating system. The ground floor, at least, should have underfloor heating. Heat rises and so upstairs rooms can be heated using radiators.

Figure 10.1 Rising energy costs

An energy consultant should be consulted before installing a heat pump system in existing building stock.

The following observations show that it is always worthwhile installing a heat pump, especially as energy costs are certain to rise further, as shown in Figure 10.1.

Potential customers are often put off by the high initial costs, especially for a brine-water heat pump system with borehole heat exchangers.

But do not forget that tapping a heat source (whether a well, borehole or ground heat exchangers) is a one-off cost and an investment in the future.

And the US$200 a barrel mark will soon be reached!

The lowest heating costs compared to all conventional heating systems:

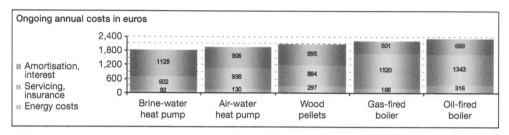

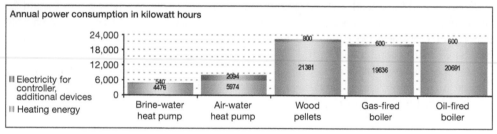

Figure 10.2 Comparing the heating costs of various heating systems

Is a heat pump economically worthwhile?

We compare various heat pump systems with gas and oil-firing heating with a solar system for a living space of 170 m² requiring heating of 10 kW, 12,000 kWh/a:

- Note: due to the requirement for increased levels of insulation in new builds, the share of the overall heat output used for generating domestic hot water is growing: it is currently around 20 per cent to 25 per cent for a simple detached home. This makes an efficient solar system essential. After amendments to the 2009 Energy Savings Ordinance (EnEV), we can assume that a solar system with supporting heating will be required, or other supplementary measures.

These simple observations show that there is nothing better than a heat pump for both new builds and renovations.

Ecological = economical! Maximum comfort with minimal energy consumption.

Table 10.1 Comparing the costs of various heating systems[a]

Investment costs	Water-water heat pump	Heat pump with borehole heat exchangers	Heat pump with ground heat exchangers	Gas-fired heating	Oil-fired heating
Heat pump/boiler	€8,250.00	€7,050.00	€7,050.00	€3,200.00	€2,900.00
6 m² solar collectors*	–	–	–	€3,500.00	€3,500.00
Well	€2,500.00	–	–	–	–
Underwater pump	€800.00	–	–	–	–
Borehole/ground heat exchangers	–	€10,000.00	€2,500.00	–	–
Brine circulation pump	–	€800.00	–	–	–
Exhaust system leading out of basement	–	–	–	€2,600.00	€2,100.00
Electrical connections including meter	€450.00	€450.00	€450.00	–	–
Gas connection/oil tank + tank room	–	–	–	€1,500.00	€12,500.00**
Circulation pump and accessories	€1,100.00	€1,100.00	€1,100.00	€800.00	€800.00
Hot water tank	€1,000.00	€1,000.00	€1,000.00	€1,200.00	€1,000.00
Buffer tank	€600.00	€600.00	€600.00	–	–
Extra cost of underfloor heating	€1,700.00	€1,700.00	€1,700.00	€900.00	–
Piping, insulation, etc.	€1,000.00	€1,000.00	€1,000.00	€1,500.00	€1,500.00
Installation costs (80 hrs)	€4,000.00	€4,000.00	€4,000.00	€4,000.00	€4,000.00
Total	€21,400.00	€27,700.00	€19,400.00	€19,200.00	€28,300.00
Annual operating costs					
Servicing/a	–	–	–	€90.00	€200.00
Electricity/a	€463.00	€495.00	€506.00	–	–
Gas/a	–	–	–	€850.00	–
Oil/a	–	–	–	–	€700.00
Chimney sweep	–	–	–	€80.00	€140.00
Total annual costs	€463.00	€495.00	€506.00	€1,020.00	€1,040.00
Seasonal performance factor	4.6	4.3	4.2	–	–

* The German renewable energy act (EEWärmeG) that came into force on 1 January 2009 requires a percentage of a building's heating requirements to be covered using renewable energy:
- For solar power in combination with a gas or oil-fired boiler at least 15 per cent.
- For gaseous biomass at least 30 per cent.
- For liquid or solid biomass at least 50 per cent.
- For geothermal energy or ambient heat at least 50 per cent –only a heat pump can do this!

** High costs for the oil tank, including the tank room.

Note: a Errors and omissions excepted. Customers are encouraged to check these details and adapt accordingly. Status: January 2009.

The economic efficiency is demonstrated as follows:

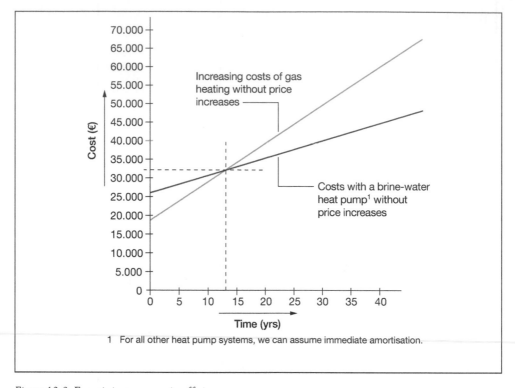

Figure 10.3 Examining economic efficiency

This illustration shows that the most expensive heat pump system using borehole heat exchangers will have amortised after around 15 years. For all other heat pump systems, we can effectively assume almost immediate amortisation.

The illustration compares the costs and amortisation of a gas-fired boiler plus solar system (as required by the EEWärmeG) with those of a brine-water heat pump feed via borehole heat exchangers. It does not account for interest rates and rising energy costs. With our current low interest rates but more strongly rising energy prices, we can assume a significantly shorter amortisation period.

Therefore, the heat pump is clearly the most economical heating system for future new builds, as well as for renovating existing buildings where they are suitable.

After the amendments to the EnEV 2009, permitted transmission heat losses will be reduced by up to a further 30 per cent. Consequently, the overall share of the heat required for hot water generation will rise, to around 25 per cent to 30 per cent for a simple detached home. This means that a solar system will need to support the heating system, with gas and oil-fired boilers becoming a thing of the past for new builds.

The illustration clearly shows the annual energy and cost savings, particularly after amortisation. Added to this is the luxury of cooling, which a heat pump system offers almost free of charge.

Similar diagrams can also be drawn up to compare other factors.

10.2 Potential savings for municipalities and regions

Heat pumps can also offer economic benefits for municipalities. A call to urban planners: Save costs and the taxpayers' money you are pumping into complicated and expensive gas networks, because heat pumps do not need gas! If a new residential area is not connected to the gas network, most homeowners will usually turn to renewables anyway. No one will bother considering installing an oil-fired heating system, which is already more expensive than any heat pump, quite apart from the higher subsequent costs. Consequently, it only makes sense to use renewables, including solar power, wood pellets and geothermal energy. And as heat pumps simply offer more benefits, most housebuilders will choose a heat pump.

This should not make any difference to urban utilities, for instead of selling gas they sell electricity. And added to this, electricity is safer than imported gas, because what happens when gas-exporting countries suddenly turn off the gas tap, needing the gas themselves? We can generate sufficient quantities of electricity here in Germany, with an energy mix including solar power and wind power, as well as hydroelectric power stations, etc. This is environmental protection in practice!

10.3 Considering amortisation rates for various forms of heating

Here, we look at various types of heating system.

In urban areas where municipalities have access to an extensive gas supply network, there is a high number of gas heating systems. Rural areas tend to use more oil-fired heating systems, and here any gas heating can only be run using liquid gas.

We also have coal-fired power stations, especially in former mining areas.

Heating systems using pellets or wood-fired boilers are also becoming increasingly popular.

As it only makes sense to compare installation costs for new builds, and our observations above have demonstrated that gas and oil-fired heating systems no longer make economic sense, we will no longer consider gas or oil.

Now we look at the costs of heating using wood pellets:

Pellet boilers

First, we need to calculate the cost of installing a pellet boiler for the required heating load:

Pellet boiler	€ _____
Stove	€ _____
Conveyor feeder	€ _____
Space for storing pellets	€ _____
Pellet storage, including piping for the filling system	€ _____
Hot water tank	€ _____
Buffer tank	€ _____
Heating pump	€ _____
Switching valve	€ _____
Expansion tank, complete	€ _____
Safety assembly	€ _____
Piping and installation	€ _____
Total costs:	€ _____

Now we need to determine the operating costs, based on the annual heating requirement:

Pellets with approx. _____ kWh/Kilo × _____ kWh × _____ €/kg = € _____
Delivery charges   € _____
Servicing costs, twice a year   € _____
Chimney sweep   € _____

Total costs: € _____

Here, too, it is clearly worth considering a heat pump!
    When we take into account the installation costs, a complete pellet-based heating system is comparable to the costs of an oil-fired system.
    Other methods of heating, such as gas and oil, as well as night storage heaters, district or local heating systems, etc. can be compared in this manner.
    Naturally, a heat pump cannot be installed in every house. A heat pump requires certain fundamental conditions, especially the ability to heat the house with low flow temperatures. In general, using a heat pump in a new build is the most comfortable and sensible solution, because here it is always possible to use surface heating.
    Heating with pellets is certainly a good solution for existing buildings. Alternatively, the complete heating system can be overhauled, allowing it to be heated using low flow temperatures, and therefore a heat pump.
    For purely economic reasons, increasing energy prices are stimulating growing demand for heat pump systems:

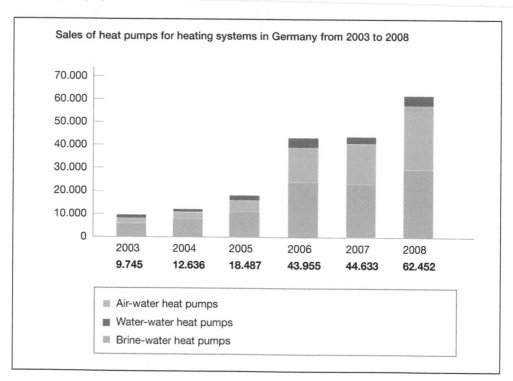

Figure 10.4 Growing demand for heat pumps

It is clear that the growing sales of heat pumps are running roughly parallel with increasing energy prices.

However, those who are clever rely on nature, for there is sufficient geothermal energy available for us all.

Figure 10.5 A GeoMax® heat pump

10.4 Cheap or good? I cannot afford to buy cheap!

The following calculation should serve to demonstrate that it does not make sense to install a 'cheap' heat pump system.

Heat pump systems can be offered cheaply if:

- The output is lower – with an electric heating element on hand as a backup.

  - The heat pump is smaller and cheaper: savings approx.   €400
  - The borehole heat exchangers are   savings approx.   €500
    smaller and cheaper:

  The consequence:

- On cold days, the heat pump cannot produce sufficient heat and/or the electric heating element is switched on. This consumes extra power, because the heating element consumes three to four times more electricity (= money) than a heat pump.

  - No buffer tank: savings approx.   €1,000

The consequence:

- The heat pump has to turn on and off far more frequently, subjecting it to far greater wear and tear. In order to counteract this, there can be no/fewer individual room thermostats so that the screed flooring can act as a buffer instead.
  - Savings from not having individual thermostats approx.   €400
  - Increased wear and tear and less comfort thanks to greater temperature fluctuations.

Repairing a compressor costs more than the savings involved!

- No hot water generation: savings of up to €1,200

The consequence:

- Domestic hot water is generated using electrical heating. This is expensive over the long-term because the heating element uses two to three times more power (= money!) than a heat pump.

  - *Immediate overall savings:*   *approx.* €3,500
  - *But later it costs you multiples of this!*

Quality is important – and it cannot be bought cheaply!

Here, a quote from John Ruskin (1819–1900) about pricing:

> There is hardly anything in the world that some man cannot make a little worse and sell a little cheaper, and the people who consider price only are this man's lawful prey.

It's unwise to pay too much, but it's worse to pay too little. When you pay too much, you lose a little money – that's all. When you pay too little, you sometimes lose everything, because the thing you bought was incapable of doing the thing it was bought to do.

The common law of business balance prohibits paying a little and getting a lot – it can't be done. If you deal with the lowest bidder, it is well to add something for the risk you run, and if you do that you will have enough to pay for something better.

11 Laws and institutions for the protection of people and the environment

A variety of guidelines and regulations have been passed to protect both people and the environment.

The emission of dangerous exhaust gases or liquids must be prevented in order to protect the environment. Consequently, only CFC-free gases may be used in refrigeration units and thus also in heat pumps. Larger refrigeration units containing more than 3 kg of refrigerant must be checked annually for leakages under operating conditions. These checks are more frequent where more than 30 kg of refrigerant is used. A safety pressure switch is prescribed to limit pressure and prevent the unintended and uncontrolled leakage of refrigerant. It also serves to protect anyone near an open refrigeration unit. Suitable measures must be taken to prevent glycol or antifreeze agents from entering the ground.

11.1 Standards and guidelines

Building laws for new builds and renovations have been enacted to consistently reduce building heat loss and encourage the use of renewable energy. There are also many guidelines and standards covering the installation and operation of heat pumps and heat pump systems. (See www.beuth.de for German standards and guidelines.)

German Energy Saving Act (EnEG)
The German Energy Saving Act (EnEG) dates back to 1977 and was the response to the oil crisis. The EnEG empowers the German federal government to pass relevant regulations and laws, and is aimed at increasing the energy efficiency of buildings. It was at this time that the Thermal Insulation Act (WSVO) was passed to reduce heat loss from buildings. In 2002, the WSVO was replaced by the EnEV 2002.

German Energy Saving Ordinance (EnEV)
The German Energy Saving Ordinance (EnEV) defines heat loss limits for new builds and renovations. It takes into account the use of renewables.

The new EnEV came into force in October 2009. Its aim is to reduce transmission heat losses by a further 30 per cent compared to the former EnEV. However, heat loss in a building includes heat loss through ventilation, and this cannot be compensated by better insulation. Therefore, the heat consumption for a new build or renovation is around 15 per cent to 20 per cent below the requirements of the earlier EnEV 2007. Anyone now building or renovating in Germany must adhere to the EnEV, and this is achievable with a well-insulated and tight building shell. An efficient heating system designed to meet the needs of the residents is also an important factor in meeting these requirements. The new

EnEV evaluates both the lowest possible heat consumption and the effectiveness of the heating system as a whole.

Currently, 40 W/m² to 45 W/m² is used for rough calculations of the heating load needed to heat a building. This value can be used to obtain an initial quote for a heating system for a new build, where the new build meets minimum EnEV standards. Where extra insulation measures result in a reduced heating load, then this must of course also be taken into account and the heating system planned correspondingly.

PLEASE NOTE

A rough calculation of heating load is never a substitute for a precise calculation of the heating load in accordance with DIN EN 12831!

In this book, heating loads are not calculated in accordance with DIN EN 12831 and therefore they should be regarded only as rough guidelines.

Furthermore, the amended EnEV 2012 will reduce acceptable heat loss limits by a further 30 per cent.

German Renewable Energy Heat Act (EEWärmeG)

The new EnEV came into force on 1 January 2009, and is based on the German Renewable Energy Heat Act (EEWärmeG). The EEWärmeG stipulates that a share of the energy required for heating must come from renewable sources; here, a well-planned heat pump system has a role to play.

The following regulations currently determine the use of renewables in Germany:

- For solar power, at least 15 per cent of the energy needed for heating must be derived from solar power.
- For gaseous biomass, at least 30 per cent of the heating load must be covered by renewables.
- For liquid or solid biomass, at least 50 per cent of the heating load must be covered by renewables.
- For geothermal energy or ambient heat, at least 50 per cent of the heating load must be covered by renewables. This is easily achieved using a good heat pump system.

Where comparable measures have the same effect of reducing the use of fossil fuels and consequently CO_2 emissions, these are also permitted as an alternative to renewables.

In principle, the new EEWärmeG applies to all new builds. However, each German federal state also has the option of making the EEWärmeG applicable for existing building stock (as is the case in Baden-Württemberg).

The EEWärmeG continues to determine eligibility for public funding.

VDI 4640

The VDI guideline regulates the thermal use of the ground and covers the following:

- Part 1: Fundamentals, authorisations, environmental aspects
- Part 2: Ground-coupled heat pump systems
- Part 3: Subterranean thermal energy storage
- Part 4: Direct use

This guideline applies to all heat pump systems with ground coupling because all heat pump systems in contact with the earth represent a potential danger to the soil. An accident could result in groundwater being contaminated with dangerous substances. This guideline covers:

- water-water heat pump systems;
- brine-water heat pump systems with borehole heat exchangers;
- brine-water heat pump systems with ground heat exchangers;
- heat pump systems with direct evaporators; and
- heat pump systems with foundation piles (energy piles).

This guideline also covers how the systems are connected, via the manifold circuit, connecting pipes and fixtures, as well as the heating system itself. It also covers the closing off and dismantling of components installed in the ground when a heat pump system is decommissioned.

VDI 4650

This guideline specifies a simplified process for determining the seasonal performance factor (SPF) for heat pumps used for heating. It does not currently cover hot water generation. This VDI guideline is also being currently reworked.

DIN EN 378

This European standard describes the technical safety mechanisms and environmental requirements for refrigeration units and heat pumps, with particular regard to checking that there are no leakages during manufacturing and during any repairs, etc.

DIN 8901

This European standard is designed to protect the ground as well as groundwater and surface water.

It states that a pressure monitor is obligatory for all brine-water heat pump systems, and this monitor must shut down the heat pump when the pressure falls below a minimum level.

Author's note:

According to DIN 8901, the pressure monitor will only turn off the heat pump if there is a leak in the brine circuit. This will not prevent further glycol from leaking into the ground – something that can only be achieved if a patented Geo-Protector has been installed.

DIN EN 12263

This standard describes the internal safety switching mechanisms required for limiting pressure (high pressure switch) within heat pumps (refrigeration units) and how they should be checked.

DIN EN 12831

This European standard regulates the calculation of heating load to obtain a precise calculation of required heat output, taking into consideration all characteristics specific to the user and the building.

The standard external temperatures are also laid down in DIN EN 12831 Supplement 1.
This standard applies when planning all heating systems.

DIN EN 14511

This European standard (sections 1 to 4) describes air conditioners, liquid cooling units and heat pumps with electrically driven compression that are used for heating and cooling rooms.

- Part 1 defines the terms.
- Part 2 determines the testing conditions and describes the standard reference conditions.
- Part 3 determines the testing process.
- Part 4 sets the minimum specifications for electrical heat pumps for heating and cooling rooms.

German Water Resources Act (WHG)

Authorisation procedures are determined by German Water Resources Act (WHG) together with the water laws applicable in each German federal state and the relevant administrative regulations:

- Permission according to § 7 (wHG)
- Approval according to § 8 (wHG)
- Refusal according to § 6 (wHG)

German Federal Mining Act (BbergG)

According to § 3, paragraph 3, no. 2, letter b of the German Federal Mining Act (BbergG), geothermal energy is a non-mined natural resource. The BbergG is applicable at depths of 100 m and more where procedures must be undertaken in accordance with mining law. Advanced notice of operations must be given.

German Carbon Chemicals Regulation (ChemKlimaschutzV) of 2 July 2008

This regulation is designed to protect the climate from changes caused by certain fluorinated greenhouse gases.

Ozone Layer Chemicals Order (ChemOzonschichtV) of 12 October 2006

This regulation protects the climate from changes caused by refrigerant gases entering the atmosphere. It includes regulations on chemicals and waste that are aimed at reducing emissions of substances that damage the ozone layer.

For refrigerators, and therefore also for heat pumps, it specifies the following servicing intervals, depending upon the volume of refrigerant used:

> 3 kg refrigerant: every 12 months
> 30 kg refrigerant: every six months
> 300 kg refrigerant: every three months

During servicing, a refrigeration engineer must check the cooling cycle for leakages.

11.2 Protection of the ground, groundwater and surface water

European standard DIN 8901 is the relevant standard when it comes to protecting the ground as well as groundwater and surface water. It requires the installation of a pressure monitor that turns off the heat pump when the pressure falls below a minimum level.

The aim is to minimise environmental damage in the event of a leak. However, it does not prevent the possibility of larger volumes of glycol leaking into the ground.

As more and more heat pumps are installed, and as existing installations start to age, leaks become increasingly likely. Therefore, the challenge is to increase groundwater protection, an important public resource. We must therefore assume that material faults, incorrect installation, incorrect materials and general material aging will all result in more frequent leakages. Ageing grouting, material faults in the deflecting heads of the borehole heat exchangers, etc. can lead to leaks. If, for example, the wrong materials are used for grouting, then they can be broken down by CO_2 in the groundwater, potentially leading to leaks.

Groundwater protection can be effectively increased by installing a Geo-Protector (see Section 6.2.2, 'The function of the Geo-Protector').

12 Starting up heat pump systems

There are several important things to consider when starting up a newly installed heat pump system for the first time. We outline a few key considerations below.

In principle, the manufacturer's instructions must be followed when starting up a heat pump. It may often be a good idea to have one of the manufacturer's own technicians start the heat pump for the first time. The technician can then also programme the controller in accordance with the customer's own requirements, the type of building and the particular model of heat pump.

The entire heat pump system should be examined prior to start-up to ensure it has been properly installed. Check the following:

- Are the hydraulic connections for the heat pump, tank, pumps and any switching valves correctly installed in accordance with the plans?
- Is the primary pump (water, brine) the right size? Is the primary pump feeding the heat pump correctly? It is important not to confuse feed and return flows.
- Are the feed and return flows correctly connected to the heat pump on the heating side?
- Is (are) the secondary pump(s) (heating pump(s)) correctly dimensioned?
- Is the switching valve connected correctly? Note the manufacturer's instructions.

Prior to switching on the heat pump, the electrical installations should be also checked, as follows:

- Is the power supply suitably fused?
  If no soft starter is installed, then the time lapse on the fuses must be sufficiently generous so that the high starting current does not trip the safety mechanism.
- Are all the three-phase connections correct, especially the direction of rotation?
- Has the controller been correctly programmed?
- Has the utility blocking been applied?
- Are all the pump and switching valves correctly connected?
- Are all the monitoring sensors correctly connected and positioned?
  Incorrectly positioned sensors can lead to faults (often high pressure faults). The sensor on the hot water tank is often placed too low. When installing the sensor, ensure that it is securely mounted – it is not sufficient to loosely place the sensor between the tank and the insulation.

TIP

For those who have little or no experience with a heat pump, it is recommended that the manufacturer's own technicians start up the heat pump system in order to avoid any damage caused as the result of incorrect settings.

Customer service technicians know the controller best and can ensure the settings meet the customer's requirements, the type of building and model of heat pump.

Prior to starting up the heat pump system, the installation on the heating side should also be complete and vented.
Incomplete or faulty installation can lead to costly delays and/or additional work – all of which is avoidable if the installation has been properly planned.

12.1 Starting up a water-water heat pump system

There are a couple of extra points to be considered when starting up a water-water heat pump for the first time. It is very important that the primary pump, usually an underwater pump, is sufficiently large and properly connected. If the evaporator in the water-water heat pump is not supplied with sufficient quantities of water, then there is a danger that it will freeze. This leads to expensive repairs. Freezing can deform the water channels, even causing ruptures.

Before being turned on, check all the electrical connections, including the fuses.

The primary pump (underwater pump or centrifugal pump) should be switched on sufficiently far in advance to ensure that enough water will flow through the evaporator when the compressor is turned on.

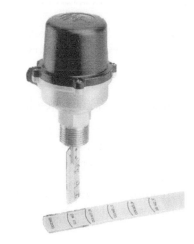

Figure 12.1
Frozen channels in the evaporator; the illustration shows the consequences of freezing; the rounded deformations caused by freezing can be clearly seen in the lower right of the photo

Figure 12.2
Flow monitor

A water-water heat pump also needs a flow monitor that immediately turns off the heat pump if insufficient quantities of water flow through the heat pump evaporator or compressor. Figure 12.2 shows a flow monitor. The flow monitor must be correctly adjusted so that it turns off the heat pump in good time.

If water levels in the well are very low, it makes sense to install dry running protection for the underwater pump. This will shut down the underwater pump before it runs dry. When correctly installed, it can also turn off the heat pump. This protects both the underwater pump and the heat pump from the consequences of insufficient water flows.

12.2 Starting up a brine-water heat pump system

There is also a danger of freezing when starting up a brine-water heat pump for the first time!

Therefore, the following should be checked prior to start-up:

- Are the borehole heat exchangers filled with a water-glycol mixture or only with water? Before starting up, the water-glycol mixture must be as homogenous as possible. Too little glycol in the mixture can lead to water freezing in the evaporator.
- Are the borehole heat exchangers correctly connected and vented?
  Please note that it is very difficult to remove air bubbles when venting borehole heat exchangers. Pressure of at least 8 bar and a correspondingly high pumping rate is needed to force an air bubble up through a borehole heat exchanger with a depth of 80 m. Therefore, where possible, all borehole loops should be individually vented. However, this is only possible in systems with a brine circuit manifold that allows the loops to be individually closed off and flushed through. Service stations are recommended for rinsing and venting. If it is not possible to rinse and vent borehole heat exchanger loops one by one, then alternating the flow direction removes the air bubbles from the borehole heat exchangers on both sides by rinsing them alternately.

A pressure monitor is recommended to monitor the brine circuit and turn off the heat pump when the pressure falls below a minimum preset level (indicating the presence of a leak).

12.3 Starting up an air-water heat pump system

Starting up an air-water heat pump is relatively simple; often only the points explained in Section 5.5.3, 'Example of planning an air-water heat pump system', need to be considered.

If using a compact unit, then make sure the cold exhaust air cannot mix with the incoming fresh air.

For a split unit (evaporator installed outside, heat pump in the basement), the connection for the refrigerant needs to be examined to ensure that it will not leak.

12.4 Drying out and warming up

As a result of increased levels of insulation in buildings, less power is required for heating – a welcome development. However, in some cases, this makes it difficult to dry out a new build using a heat pump, and even impossible during some periods of the year. Here, extra electrical heating can be helpful, or a construction dryer.

Usually, the freshly laid floor screed in a new build must first be warmed up before tiles can be laid. The temperature needs to be slowly and consistently raised, maintained, and then slowly and steadily lowered. Only once this process has been completed can further work on the floor proceed. This floor screed warming function is programmed into most heat pump controllers.

13 Common heat pump errors

A heat pump system is relatively complex and not comparable to a gas or oil-fired heating system. It usually involves several technical specialisms, all of which must be coordinated. They are:

- well engineering for borehole heat exchangers or wells (except for air-water heat pump systems);
- heating engineering;
- sanitary engineering for hot water pipes; and
- electrical engineering for controlling the heat pump system and the individual electrical components.

13.1 Well drilling errors

The well engineer is responsible for drilling the intake and discharge wells.

The following points must be considered when installing a water-water heat pump:

- The well engineer should be certified (in Germany according to DVGW W120), indicating a correct level of relevant training and knowledge of the legal regulations. It is particularly important that access to the various aquifers is sealed off and that they are kept separate. If the water regulations are not adhered to, then it should come as no surprise when the local water authorities refuse to issue permission for the heat pump system to operate.
- The intake well should be drilled so that the water it contains is free of iron and manganese, in accordance with drinking water regulations. Where the statutory iron and manganese limits are not exceeded, the heat pump can safely operate using water drawn from the well.
- Attention must also be paid to the water quality and, where necessary, the appropriate materials used (e.g. a corrosion-resistant evaporator).
- The intake well must have sufficient supplies of water for long-term use.
- The discharge well must also have the capacity to absorb all the discharged water over the long-term. In general, if the discharge well can absorb the outgoing water, then the intake well will be able to supply the heat pump with sufficient quantities of water.
- The underwater pumps must be selected in accordance with the type of heat pump, well and groundwater level.

Furthermore, a long-term minimum water flow rate must be available if safe operation is to be ensured. The flow rate must be monitored to protect the heat pump from freezing.

13.2 Errors in installing borehole or ground heat exchangers

The well engineer must correctly dimension the borehole heat exchangers.
The following must be noted when planning and installing the borehole heat exchangers:

- The well engineer should be certified (see above).
- The borehole heat exchangers must be large enough for the heat pump and to heat the building. The brine temperature should not fall below 0 °C during heating. Overcooling can lead to freezing, and this in turn leads to a low pressure fault and the heat pump being shut down. When using ground heat exchangers, freezing can lead to the ground rising in the immediate area. This in turn can cause damage (e.g. where ground heat exchanger loops run close to terraces or fences, etc.).
- The borehole heat exchangers must all be installed in the same manner and with the brine loops all of the same length. This ensures that the brine flows evenly through the loops. Where this is not the case, brine circuit manifolds with flow plates must be installed.
- Each borehole loop must be connected to the manifold according to the Tichelmann principle. They can be installed by the heating engineer, if so required.

13.3 Errors in the hydraulics

The heating engineer or even the planner is responsible for the hydraulics.
The following must be considering during planning and execution:

- The primary pump (water or brine pump) must be large enough to ensure the heat pump is supplied with sufficient volumes of water or brine. If the pump is too small, this can lead to low pressure faults and the heat pump shutting down. It is vital that sufficient volumes of water flow through a water-water heat pump to prevent the evaporator from freezing. Safety mechanisms such as flow monitors and dry running protection are recommended.
- The secondary pump(s) must also be large enough to ensure that the heat pump can transfer the heat it produces. Where this is not the case, a high pressure fault will cause the heat pump to be shut down.
- The heat exchanger in the hot water tank must be large enough to ensure that the heat produced by the heat pump can be transferred. Where this is not the case, again a high pressure fault will cause the heat pump to shut down.
- Where the heat pump has no buffer tank (not recommended), steps must be taken to ensure that the heat produced by the heat pump can be taken up. Here again, where the heat is not sufficiently absorbed, a high pressure fault will lead to the heat pump being shut down.
- The hot water piping should be large enough to ensure 'hot water comfort'.

13.4 Installation errors

The following is a list of possible installation errors and their causes:

- Using a combined storage tank/also using a buffer tank for hot water generation. Combined storage tanks are excellent for use with a solar system. Depending on the quantity of solar power generated, the controller ensures that the combined tank or buffer tank is charged at the correct point.

  However, this application makes no sense for use with heat pumps. Why? In heat pump systems, the water used for heating must be kept at a flow temperature that is as low as possible yet hot enough for heating. This is not possible using standard buffer or combined storage tanks.

  As a heat pump can only heat the returning flow by around 5 K, these tanks will automatically heat all the water in the tank to the temperature required for hot water generation. Consequently, it is not possible to layer water of different temperatures when using a heat pump. As a result, the overall level of efficiency can only be unsatisfactory.

- The sensor is incorrectly positioned on the hot water tank.

  If the sensor is placed too high, then the heat pump will not properly heat the water in the tank, leading to insufficient quantities of hot water being available.

  If the sensor is positioned too low on the hot water tank, then the heat pump tries to heat all the water in the tank to the specified temperature. If the hot water tank has been heated right through but the set temperature has not yet been reached at the point at which the sensor is mounted, then the heat generated by the heat pump can no longer be given off. This leads to a high pressure fault.

- The sensor is incorrectly positioned on the buffer tank.

  If the sensor is positioned too high, then the buffer tank cannot be charged properly. Only the upper levels of the tank are heated. Therefore, the heat pump is shut down before the buffer tank is completely heated through, and therefore not much heat is buffered.

  Consequently, the heat pump turns off too early, and this leads to it being switched on and off too frequently. A key advantage of having a buffer tank is lost.

- The sensor has become detached.

  This can occur when the sensor has not been properly fixed between the insulation and the tank, and therefore it slips. We need to assume that the insulation expands more than the tank itself when the tank heats up. A loosely placed sensor can then slip. In extreme cases, contact with the tank can be completely lost. If this happens, the heat pump controller no longer recognises when the tank is fully heated (because the sensor is measuring a different, lower temperature) and tries to continue heating. Eventually, this results in a high pressure fault and the heat pump is turned off.

13.5 Error messages and their possible causes

The following is a list of the most common error messages and their causes:

Low pressure fault
The heat pump is not being fed with sufficient quantities of heat from the heat source, because:

- the underwater pump or brine pump is too small or faulty;
- the borehole or ground heat exchangers are too small; or
- the incoming flow (filter) is blocked (this should trip the motor fuse on the primary pump).

Refrigerant has escaped from the heat pump, because:

- there is a leak in the refrigeration circuit; or
- the evaporator is faulty.

High pressure fault
Insufficient heat is being transferred away from the heat pump, because:

- the charging pump is too small;
- the charging pump is faulty;
- the switching valve is faulty;
- the switching valve switches too slowly and so breaks/reduces the primary circuit for too long;
- the heat exchanger in the hot water tank is too small;
- the hot water temperature is set too high;
- the heat pump is operating without a buffer tank and there is too little heat transfer;
- the hot water tank is too small; or
- the sensor has been positioned too low on the hot water tank.

Fault in the safety chain
Depending on the type of heat pump or heat pump controller, there are one or more safety inputs. These monitor the inputs, either individually or collectively. These safety chains include tripping fuses, motor circuit breakers (compressor and underwater or brine pumps), etc. All auxiliary switches must be designed as openers (i.e. one of the auxiliary contacts opens in the event of a fault). This is registered as a line break and will not render the safety chain inoperative.

Fault in the flow monitor, dry running protection
These faults are particularly relevant in water-water heat pump systems. The flow monitor ensures that the heat pump is fed with sufficient volumes of water, shutting off the heat pump if the flow rate is too low. The dry running protection is designed to protect the underwater pump; it is not a replacement for the flow monitor because it stops working if the underwater pump fails.

13.6 Operating faults

The operator can also cause errors through incorrect use or faulty settings with the consequence that a building is insufficiently heated. The following are a few examples:
The heat output and flow temperatures of a heat pump are usually adjusted to meet the needs of the building in which it is installed.

The heat output of a heat pump should not be too high. If the heat pump is too large, this can involve greater overall costs, especially for brine-water heat pumps for which borehole or ground heat exchangers must be installed.

Most controllers include a reduced operational setting, known as a night mode. The operator is often familiar with these settings from previous heating systems. With the aim of saving energy and heating costs in the usual manner, the operator programmes the heat pump to include periods of reduced operation. These are normally at night (i.e. when the external temperature is lowest). As heat is continually lost via the exterior of the building, the building cools down. If well insulated, however, this cooling is not extreme enough for the actual value to fall below that of the set night mode value. On the other hand, this also means the heat pump supplies no heat to the building during these off periods. An off period from 10 p.m. to 6 a.m. represents a downtime of eight hours! On cold days, this can be disastrous; the remaining 16 hours of the day may be insufficient to heat the building adequately. And if off periods are also programmed for periods of absence, then the operator should not be surprised when he cannot get the house warm enough on cold days.

The flow temperature should be as low as possible – this is the only means of ensuring that the heat pump system as a whole can operate efficiently. Where the flow temperature is too high, then the efficiency level falls. Consequently, the heating circuit also operates using as low a flow temperature as possible and, in turn, this requires that all rooms be heated to as similar a level as possible. However, if the operator lowers the thermostat setting too far in several of the rooms, then these rooms may receive no heat from the (usually) underfloor heating. As there is no insulation between the interior walls, heat flows from the heated to the unheated rooms. However, the heat output in the heated rooms is not sufficient to heat both heated and unheated rooms. The result is heated rooms that are not warm enough, and unheated rooms that remain cooler (i.e. the heated surface area/output of the heated rooms is not enough to heat all the rooms, including those in which the thermostat has been turned right down).

The operator may set the heating curve on the controller incorrectly. Where this setting is incorrect (base point, steepness of the rise, heating limit, etc.), the building may not be properly heated. The flow temperature is either too high, with a negative impact on the system's efficiency, or too low, so that the building is insufficiently heated. This is often noticeable during the transitional periods in autumn and/or spring. A qualified technician should carry out the necessary adjustments, especially for newly installed systems.
The operator of the heat pump system should be fully informed of this eventuality, or a qualified technician should offer to readjust the system.

14 Concluding observations and outlook

To date, heat pumps are the most modern and efficient means of heating available, and so demand for heat pumps is growing. A heat pump is ecological because it uses geothermal energy, provided by the sun. A heat pump operates in the most ecological manner possible when 'green electricity' is used to drive the pump. Heat pumps are also economical because they involve the lowest operating costs. Heat pumps also offer the added luxury of cooling (i.e. a heat pump system can both heat and cool).

In short:

Maximum comfort and environmental protection with minimal heating costs!

That is something you only get with a heat pump.

There are currently alternatives on the market, including:

- Wood heating systems (pellets, wood, etc.) yet there is not enough wood available to heat every house. Heating with wood is certainly one of the most ecologically friendly means of heating because the CO_2 cycle is closed and no additional CO_2 is pumped into the atmosphere. The disadvantage is the high levels of maintenance and the high operating and maintenance costs.
- Combined heat and power generators (CHP) – but here, 100 per cent of the energy to drive the system must be transported, and this involves transmission losses. The high level of maintenance required (like a car) must also be taken into account.
- Fuel cells – here again, 100 per cent of the energy to drive the system must be transported, involving transmission losses.

This indicates a rosy future for heat pumps, particularly in new builds!

TIP

In Germany, there are plans to introduce legislation requiring municipalities to ensure that 15 per cent or more of the heating power used in new builds is generated using renewables. For gas or oil-fired heating systems, this means adding supplementary solar systems. This effectively increases the costs of conventional heating systems, correspondingly reducing the amortisation period for heat pumps.

What if all houses were equipped with heat pumps? Would the ground become too cold?

Of course, heat would be extracted from the ground, leading to local falls in ground temperature. However, in Germany, a small modern terraced house, designed for a single family and with around 150 m² of living space, usually sits on a plot of land of around 200 m². According to the ENeV, each house would need a maximum of 9 kW for heating and hot water generation. This corresponds to a cooling capacity of 7.2 kW. For 50 W/m, two borehole heat exchangers each 72 m in depth would be necessary. The minimum distance between the heat exchangers should be 5 m. Therefore, *all* the property owners in this row of terraced housing could heat their homes using a heat pump. Seen in these terms, all future housebuilders will be able to heat their houses with heat pumps.

So heat pumps have a great future.

All houses and buildings on earth can be heated using geothermal energy!

Heat pumps can also be used for both heating and cooling. When cooling during the summer, large quantities of heat are pumped back into the ground – most of which is then available again for heating in winter.

It is also possible to centrally heat (and cool) larger residential developments very economically using heat pumps. Here, water-water heat pump systems are recommended, if circumstances permit. It may also be sensible to consider communal solutions for discharging the large volumes of return water (e.g. into a stream or drainage ditches, etc.). The regions and municipalities need to step up and remove unnecessary barriers to realising ecological projects, as well as actively encouraging them (e.g. through regulations designed to make their installation both possible and easier).

What if the water contains iron or manganese, something to be expected in larger, more powerful wells?

This water can be treated underground (subterranean). The costs of such wells, including water treatment, are usually still lower than installing large fields of borehole heat exchangers, and water-water heat pumps are also around 20 per cent more efficient (COP) than their brine-water equivalents.

Such a heat pump system is shown in Section 5.3.1.4.

It would also be worth considering shared access to a single water source for a residential development, with each house being individually heated with its own heat pump. A good example of this has been demonstrated in Dorstenwulfen, in Germany. (See 'Examples of heat pumps').

The options are endless, and will only increase in future.

TIP

Exploding energy prices mean it is increasingly important for planners, engineers and architects to advise and encourage housebuilders to use the most efficient means of heating.

Intelligent minds are needed to come up with innovative solutions.

15 Exercises

The following exercises are designed as an aid to greater understanding. They should initially be solved using only the information and examples given in this book. The solutions to the exercises are intended only as a comparison and not as a code of practice. The exercises can all be solved using the information provided in each section of the book.

The following data sheets and diagrams should be used to complete the exercises:

Tables 15.1 and 15.2 Technical data for GeoMax® water-water heat pumps

Geo-Max®	WW6.400	WW8.300	WW9.800	WW11.600	WW14.200	WW17.500	WW20.100	WW23.500
Heat output								
W10 / W35	6,4 kW	8,3 kW	9,8 kW	11,6 kW	14,2 kW	17,5 kW	20,1 kW	23,5 kW
W10 / W55	5,9 kW	7,6 kW	9,0 kW	10,5 kW	13,0 kW	15,8 kW	18,8 kW	21,2 kW
COP								
W10 / W35	5,8	5,9	5,9	5,9	5,9	5,9	6,0	6,0
W10 / W55	3,2	3,2	3,2	3,3	3,4	3,4	3,4	3,4
Electrical output								
W10 / W35	1,1 kW	1,4 kW	1,7 kW	2,0 kW	2,4 kW	3,0 kW	3,5 kW	3,9 kW
W10 / W55	1,8 kW	2,4 kW	2,8 kW	3,2 kW	3,9 kW	4,7 kW	5,5 kW	6,3 kW
Nominal voltage	400 V, 50 Hz							
Nominal current								
W10 / W35	2,3 A	3,0 A	3,6 A	4,1 A	5,2 A	6,7 A	7,7 A	8,5 A
W10 / W55	3,3 A	4,1 A	5,0 A	5,7 A	7,0 A	8,6 A	9,6 A	11,4 A
Starting current	24 A	32 A	40 A	46 A	50 A	65 A	74 A	101 A
Refrigerant	R 407 C							
Evaporator								
Material	Stainless steel, copper or nickel-plated brazed							
Min. flow	1,5 m³/h	2,0 m³/h	2,3 m³/h	2,7 m³/h	3,4 m³/h	4,1 m³/h	4,7 m³/h	5,6 m³/h
Pressure loss	0,14 bar	0,23 bar	0,14 bar	0,20 bar	0,17 bar	0,26 bar	0,22 bar	0,22 bar
Connector	1"	1"	1"	1"	1"	1"		
Condenser								
Material	Stainless steel, copper-plated							
Min. flow	1,1 m³/h	1,4 m³/h	1,7 m³/h	2,0 m³/h	2,4 m³/h	3,0 m³/h	3,5 m³/h	4,0 m³/h
Pressure loss	0,07 bar	0,12 bar	0,16 bar	0,10 bar	0,15 bar	0,13 bar	0,18 bar	0,16 bar
Connector	1"	1"	1"	1"	1"	1"	1"	1"

Tables 15.1 and 15.2 Technical data for GeoMax® water-water heat pumps – *continued*

Geo-Max®	WW 26.300	WW 31.800	WW 36.500	WW 45.200	WW 55.100
Heat output					
W10 / W35	26,3 kW	31,8 kW	36,5 kW	45,2 kW	55,1 kW
W10 / W55	24,1 kW	29,0 kW	33,5 kW	41,2 kW	50,1 kW
COP					
W10 / W35	6,0	6,0	6,1	5,9	6,1
W10 / W55	3,4	3,5	3,5	3,4	3,4
Electrical output					
W10 / W35	4,4 kW	5,3 kW	6,0 kW	7,7 kW	9,1 kW
W10 / W55	7,2 kW	8,4 kW	9,7 kW	12,1 kW	14,6 kW
Nominal voltage	400 V, 50 Hz				
Nominal current					
W10 / W35	10,6 A	12,6 A	13,5 A	14,5 A	17,8 A
W10 / W55	13,5 A	15,9 A	17,1 A	20,1 A	24,5 A
Starting current	99 A	123 A	141 A	167 A	198 A
Refrigerant	R 407 C				
Evaporator					
Material	Stainless steel, copper or nickel-plated brazed				
Min. flow	6,3 m³/h	7,6 m³/h	8,7 m³/h	10,7 m³/h	13,0 m³/h
Pressure loss	0,28 bar	0,31 bar	0,32 bar	0,34 bar	0,35 bar
Connector	1¼"	1¼"	1¼"	1½"	1½"
Condenser					
Material	Stainless steel, copper-plated				
Min. flow	4,5 m³/h	5,5 m³/h	6,3 m³/h	7,8 m³/h	9,5 m³/h
Pressure loss	0,20 bar	0,20 bar	0,21 bar	0,25 bar	0,28 bar
Connector	1¼"	1¼"	1¼"	1½"	1½"

Tables 15.3 and 15.4 Technical data for GeoMax® brine-water heat pumps

Geo-Max®	SW 4.600	SW 5.900	SW 7.000	SW 8.300	SW 10.300	SW 12.400	SW 16.000	SW 16.800
Heat output								
B0 / W35	4,6 kW	5,9 kW	7,0 kW	8,3 kW	10,3 kW	12,4 kW	15,0 kW	16,8 kW
B0 / W55	4,4 kW	5,7 kW	6,7 kW	7,9 kW	9,8 kW	11,6 kW	13,9 kW	15,8 kW
COP								
B0 / W35	4,1	4,1	4,2	4,2	4,2	4,2	4,4	4,4
B0 / W55	2,4	2,4	2,4	2,4	2,5	2,5	2,6	2,5
Electrical output								
B0 / W35	1,1 kW	1,4 kW	1,7 kW	2,0 kW	2,5 kW	2,9 kW	3,4 kW	3,8 kW
B0 / W55	1,9 kW	2,4 kW	2,8 kW	3,3 kW	3,9 kW	4,6 kW	5,4 kW	6,3 kW
Nominal voltage	400 V, 50 Hz							
Nominal current								
B0 / W35	2,3 A	3,0 A	3,6 A	4,2 A	5,3 A	6,6 A	7,7 A	8,5 A
B0 / W55	3,3 A	4,2 A	5,0 A	5,8 A	7,0 A	8,5 A	9,5 A	11,5 A
Starting current	24 A	32 A	40 A	46 A	50 A	66 A	74 A	101 A
Refrigerant	R 407 C							
Evaporator								
Material	Stainless steel, copper-plated							
Min. flow	1,2 m³/h	1,5 m³/h	1,8 m³/h	2,0 m³/h	2,6 m³/h	3,2 m³/h	3,9 m³/h	4,4 m³/h
Pressure loss	0,10 bar	0,16 bar	0,10 bar	0,14 bar	0,20 bar	0,18 bar	0,17 bar	0,21 bar
Connector	1"	1"	1"	1"	1"	1"	1"	1"
Antifreeze	Antifogen N							
Condenser								
Material	Stainless steel, copper-plated							
Min. flow	0,8 m³/h	1,0 m³/h	1,2 m³/h	1,4 m³/h	1,8 m³/h	2,1 m³/h	2,6 m³/h	2,9 m³/h
Pressure loss	0,14 bar	0,23 bar	0,09 bar	0,12 bar	0,18 bar	0,25 bar	0,17 bar	0,21 bar
Connector	1"	1"	1"	1"	1"	1"	1"	1"

Tables 15.3 and 15.4 Technical data for GeoMax® brine-water heat pumps – *continued*

Geo-Max®	SW 18.800	SW 23.100	SW 26.800	SW 33.100	SW 40.100
Heat output					
B0 / W35	18,8 kW	23,1 kW	26,8 kW	33,1 kW	40,1 kW
B0 / W55	17,9 kW	21,9 kW	25,1 kW	31,1 kW	38,0 kW
COP					
B0 / W35	4,2	4,3	4,4	4,4	4,4
B0 / W55	2,5	2,6	2,7	2,6	2,6
Electrical output					
B0 / W35	4,5 kW	5,4 kW	6,1 kW	7,6 kW	9,1 kW
B0 / W55	7,1 kW	8,4 kW	9,5 kW	11,9 kW	14,5 kW
Nominal voltage	400 V, 50 Hz				
Nominal current					
B0 / W35	10,6 A	12,7 A	13,5 A	14,4 A	17,8 A
B0 / W55	13,5 A	15,9 A	17,0 A	20,0 A	24,4 A
Starting current	99 A	123 A	141 A	167 A	198 A
Refrigerant	R 407 C				
Evaporator					
Material	Stainless steel, copper-plated				
Min. flow	4,8 m³/h	6,0 m³/h	7,0 m³/h	8,6 m³/h	10,4 m³/h
Pressure loss	0,19 bar	0,21 bar	0,23 bar	0,28 bar	0,27 bar
Connector	1¼"	1¼"	1¼"	1½"	1½"
Antifreeze	Antifogen N – 30%				
Condenser					
Material	Stainless steel, copper-plated				
Min. flow	3,2 m³/h	4,0 m³/h	4,6 m³/h	5,7 m³/h	6,9 m³/h
Pressure loss	0,26 bar	0,23 bar	0,30 bar	0,30 bar	0,32 bar
Connector	1¼"	1¼"	1¼"	1½"	1½"

The heat exchangers for cooling have a pressure loss of P_{HEC} = 20 kPa.

Technical data for an air-water heat pump:

Table 15.5 Technical data, BESST air-water heat pump

Technical data – weight									

Series Besst – Besst/r

Reference standard UNI-EN 14511:2004

Model		1/9	2/10	3/11	4/12	5/13	6/14	7/15	8/16
Heat output[1]	KW	6.8	8.3	11.0	15.0	19.9	22.2	28.0	37.2
Power input[1]	KW	1.74	2.11	2.81	3.61	4.28	4.83	6.48	8.44
Heat output[2]	KW	5.9	7.2	9.5	13.0	16.6	18.9	23.8	31.7
Power input[2]	KW	178	2.20	2.90	3.67	4.33	4.86	6.55	8.52
Heat output[3]	KW	5.0	6.2	8.0	10.8	13.5	15.4	19.2	25.8
Power input[3]	KW	1.79	2.26	2.97	3.73	4.36	4.88	6.66	8.63
Heat output[4]	KW	6.6	8.1	10.6	14.4	18.4	21.0	26.4	34.8
Power input[4]	KW	2.14	2.68	3.50	4.49	5.23	5.96	7.84	10.18
Heat output[5]	KW	5.6	7.1	9.3	12.6	15.9	18.1	22.6	30.0
Power input[5]	KW	2.16	2.77	3.60	4.60	5.26	5.99	8.00	10.33
Cooling capacity[6]	KW	7.5	10.3	13.0	17.8	23.6	27.2	33.9	45.1
Power input[6]	KW	1.8	2.28	2.93	3.89	4.96	5.67	7.04	9.05
Cooling capacity[7]	KW	6.9	9.3	11.8	17.0	21.3	24.6	31.0	40.4
Power input[7]	KW	2.00	2.72	3.46	4.56	5.78	6.60	8.13	10.51
Cooling capacity[8]	KW	6.0	8.2	10.2	14.1	17.9	21.0	26.9	34.9
Power input[8]	KW	1.76	2.34	3.01	3.93	5.03	5.69	7.05	9.19
Cooling capacity[9]	KW	5.3	7.1	9.3	13.0	17.0	19.5	24.7	32.2
Power input[9]	KW	2.04	2.81	3.58	4.65	5.85	6.66	8.22	10.71
Nominal water output[10]	m³/h	1.0	1.2	1.6	2.2	2.8	3.2	4.1	5.4
Ø Hydraulic connections	Inches	3/4	3/4	1	1	1	1	1-1/4	1-1/4
Container capacity	Litre	16	16	36	36	57	57	70	70
Auxiliary heating*	KW	3.0	3.0	6.0	6.0	8.0	8.0	10.0	10.0
Supply voltage	V/50Hz	230 ~	230 ~	230 ~	400 3N ~	400 3N ~	400 3N ~	400 3N ~	400 3N ~
Protection class	IP	44	44	44	44	44	44	44	44
Propeller fans	No.	1	1	2	2	2	2	4	4
Nominal air suppy	m³/h	3.300	3.250	6.500	6.500	8.000	8.000	14.000	13.600
Nominal sound pressure[11]	dB(A)	58.0	58.0	62.8	62.8	61.5	61.5	63.0	63.0
R410a refrigerant	kg	1.7	1.9	2.7	3.1	4.9	5.5	8.2	9.6
Empty weight	kg	110	112	164	175	224	230	390	394

1 External air temperature +7 °C, water 35–30 °C
2 External air temperature 0 °C, water 35–30 °C
3 External air temperature –7 °C, water 35–30 °C
4 External air temperature +7 °C, water 45–40 °C
5 External air temperature +0 °C, water 45–40 °C
6 External air temperature +30 °C, water 18–23 °C
7 External air temperature +35 °C, water 18–23 °C
8 External air temperature +30 °C, water 7–12 °C
9 External air temperature +35 °C, water 7–12 °C
10 External reference air temperature 0 °C, water 35–30 °C
11 Sound pressure measured in free field conditions, 1.5 m from the fans and 1.5 m from the ground

Switching valves and their diagrams:

Table 15.6 Technical data for the ESBE three-way zone valve

Nominal diameter	K_{VS} value	Differential pressure	Nominal voltage
DN 20	5.7	1.0 bar	230 V, 50 Hz
DN 25	5.7	0.7 bar	230 V, 50 Hz

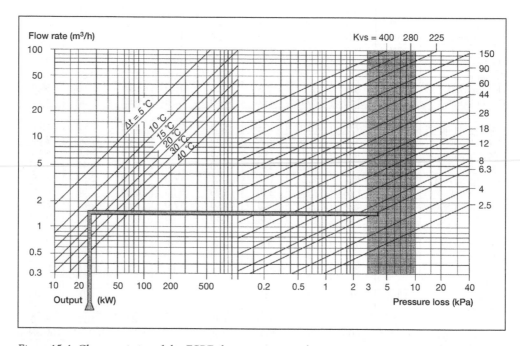

Figure 15.1 Characteristics of the ESBE three-way zone valve

Table 15.7 K$_{VS}$ values

KVS	ND	Connection	Weight (kg)
1.6	15	Rp ½″ inner thread	1.1
2.5	15	Rp ½″ inner thread	1.1
4.0	15	Rp ½″ inner thread	1.1
6.3	20	Rp ¾″ inner thread	1.3
10	25	Rp 1″ inner thread	1.5
16	32	Rp 1¼″ inner thread	2.1
25	40	Rp 1½″ inner thread	3.0
38	50	Rp 2″ inner thread	4.7

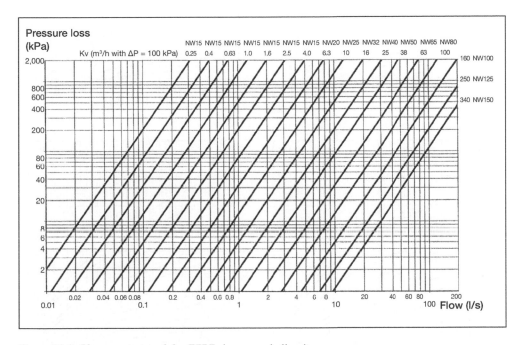

Figure 15.2 Characteristics of the ESBE three-way ball valve

For selecting the underwater pumps:

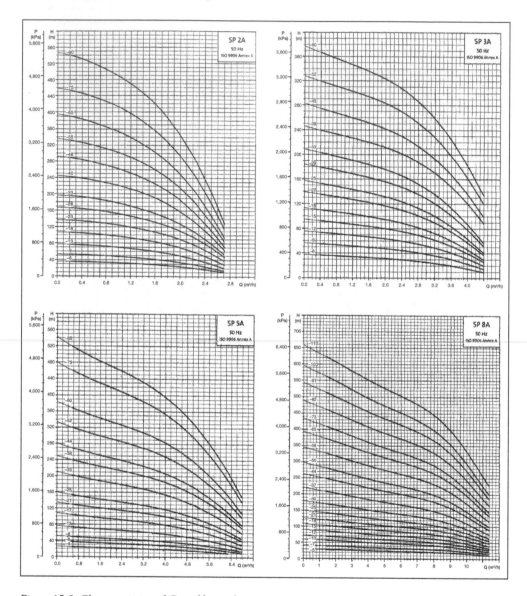

Figure 15.3 Characteristics of Grundfos underwater pumps

Pressure loss in PE pipes and moulded parts – see Section 7.1.1, 'The underwater pump and how it is monitored'.

For selecting the underwater pumps:

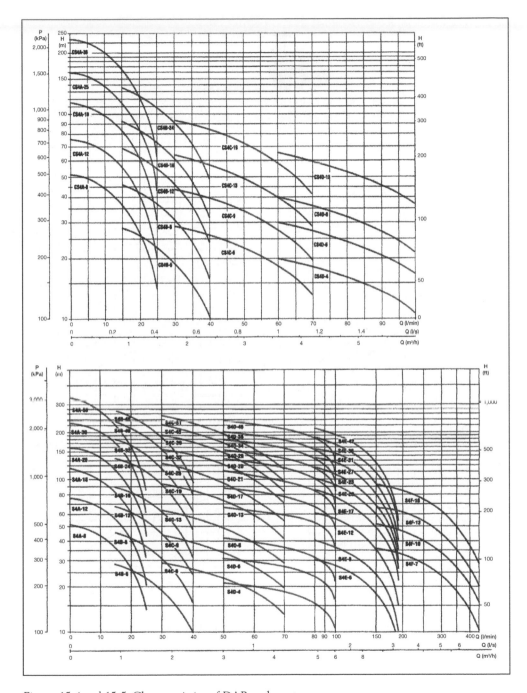

Figures 15.4 and 15.5 Characteristics of DAB underwater pumps

Pressure loss in PE pipes and moulded parts – see Section 7.1.1, 'The underwater pump and how it is monitored'.

For selecting the brine and charging pumps:

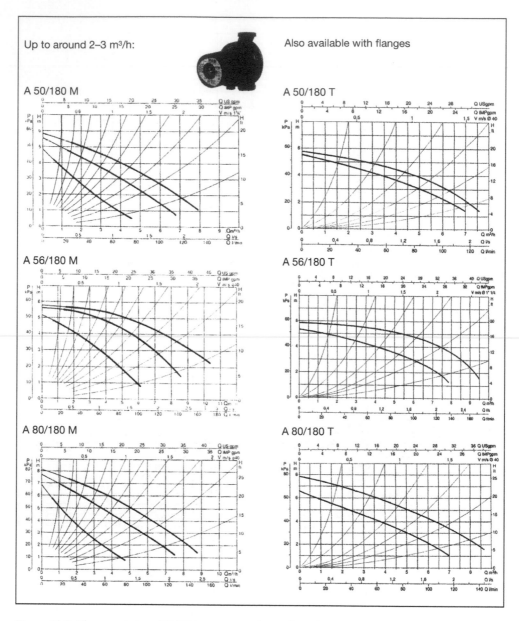

Figure 15.6 Characteristics of DAB heating circulators (circulation pumps)

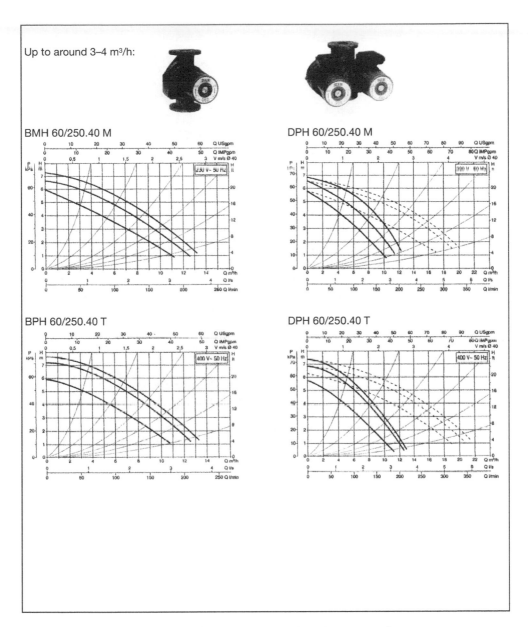

Figure 15.7 Characteristics of DAB heating circulators (circulation pumps)

For selecting the brine and charging pumps:

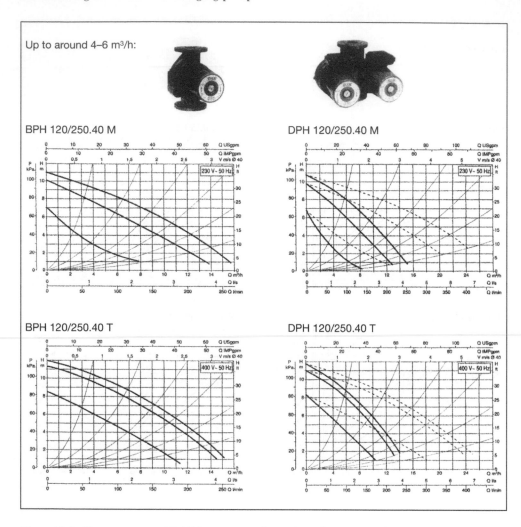

Figure 15.8 Characteristics of DAB heating circulators (circulation pumps)

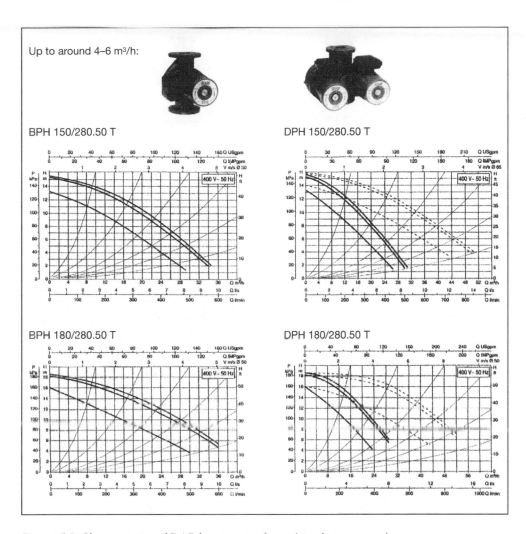

Figure 15.9 Characteristics of DAB heating circulators (circulation pumps)

For selecting the brine and charging pumps:

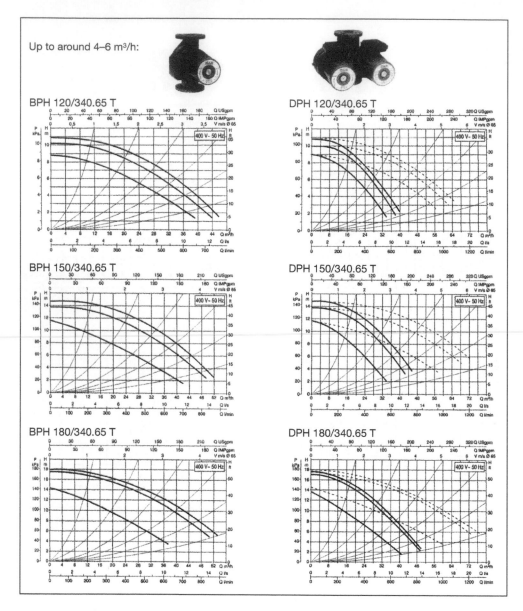

Figure 15.10 Characteristics of DAB heating circulators (circulation pumps)

To determine pressure loss in the hot water tank see Table 15.8:

Table 15.8 Technical data for the Austria Email AG bivalent hot water tank

	N_L-number	Heating surface	Throughput in kW and l/h												Flow resistance in mbar		
Flow temperature	80°C		70°C	70°C	70°C	80°C	80°C	80°C	70°C	70°C	70°C	80°C	80°C	80°C			
Hot water temperature	45°C		45°C	45°C	45°C	45°C	45°C	45°C	60°C	60°C	60°C	60°C	60°C	60°C			
Cold water temperature	10°C		10°C	10°C	10°C	10°C	10°C	10°C	10°C	10°C	10°C	10°C	10°C	10°C			
Thermostat setting	60°C																
Hot water flow	3 m³/h	m²	1 m³/h	2 m³/h	3 m³/h	1 m³/h	2 m³/h	3 m³/h	1 m³/h	2 m³/h	3 m³/h	1 m³/h	2 m³/h	3 m³/h	1 m³/h	2 m³/h	3 m³/h
BE 160 ERM	2	0,60	11,9 / 293	13,9 / 342	14,9 / 367	15,6 / 384	18,5 / 455	19,8 / 487	8,5 / 146	9,7 / 167	10,3 / 177	12,5 / 215	14,7 / 253	15,7 / 270	2	14	54
BE 200 ERM	3,5	1,00	18,0 / 443	21,7 / 534	23,5 / 578	23,3 / 573	28,4 / 699	31,0 / 763	13,2 / 227	15,5 / 267	16,6 / 286	19,1 / 329	22,9 / 394	24,8 / 427	14	45	96
BE 300 ERM	7,5	1,50	23,0 / 566	30,1 / 740	31,8 / 782	29,8 / 733	39,1 / 962	42,7 / 1050	17,1 / 294	20,9 / 360	22,4 / 386	24,8 / 427	31,0 / 534	33,9 / 584	32	90	178
BE 400 ERM	11	1,80	27,2 / 669	34,8 / 856	38,9 / 957	35,1 / 863	45,1 / 1109	50,7 / 1247	20,4 / 351	25,5 / 439	27,5 / 474	29,3 / 505	37,1 / 639	41,2 / 709	53	114	210
BE 500 ERM	15	2,00	29,8 / 733	39,2 / 964	44,2 / 1087	38,3 / 942	51,2 / 1250	58,1 / 1429	21,9 / 377	27,2 / 468	29,5 / 508	31,7 / 546	42,1 / 725	48,1 / 828	41	139	293
BE 300 ERMR below	7,5	1,50	23,0 / 566	30,1 / 740	31,8 / 782	29,8 / 733	39,1 / 962	42,7 / 1050	17,1 / 294	20,9 / 360	22,4 / 386	24,8 / 427	31,0 / 534	33,9 / 584	32	90	178
BE 300 ERMR above	1,8	1,00	16,6 / 408	20,2 / 497	21,8 / 536	21,9 / 539	26,7 / 657	29,1 / 716	12,2 / 210	14,4 / 248	15,7 / 270	18,1 / 312	21,7 / 374	23,6 / 406	20	58	121
BE 400 ERMR below	11	1,80	27,2 / 669	34,8 / 856	38,9 / 957	35,1 / 863	43,1 / 1109	50,7 / 1247	20,4 / 351	25,5 / 439	27,5 / 474	29,3 / 505	37,1 / 639	41,2 / 709	53	114	210
BE 400 ERMR above	3	1,00	16,7 / 411	20,0 / 492	21,5 / 529	21,6 / 531	26,1 / 642	28,2 / 694	12,4 / 214	14,5 / 250	15,4 / 265	18,0 / 310	21,4 / 369	23,0 / 396	12	40	83
BE 500 ERMR below	15	2,00	29,8 / 733	39,2 / 964	44,2 / 1087	38,1 / 942	51,2 / 1250	58,1 / 1429	21,9 / 377	27,2 / 468	29,5 / 508	31,7 / 546	42,1 / 725	48,1 / 828	41	139	293
BE 500 ERMR above	3,7	1,00	16,2 / 399	19,6 / 482	20,9 / 514	20,3 / 499	25,0 / 615	27,5 / 677	11,4 / 195	13,5 / 232	14,0 / 241	16,8 / 289	19,9 / 343	21,0 / 362	19	55	109

For determining pressure loss in hot water tanks:

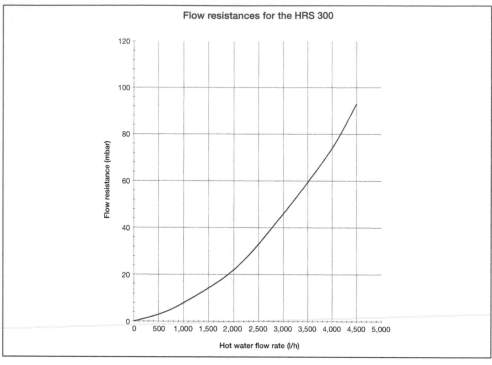

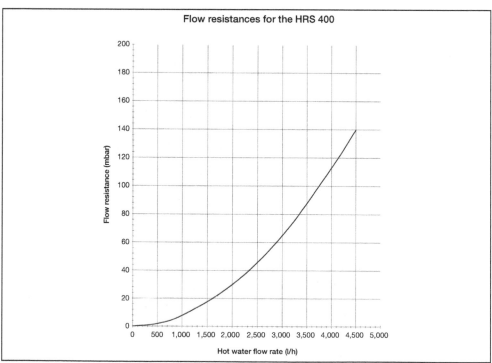

Figures 15.11 and 15.12 Characteristics of the Austria Email AG high performance hot water tank

15.1 Planning a water-water heat pump

A customer comes and asks you to plan a heat pump system for his house. Having considered the particular location and geological characteristics, you recommend a water-water heat pump. The customer does not need cooling but does want a solar system to support the heating system. The garage temperature should also be regulated in order to protect the car from corrosion. The storeroom will not be heated. The house plans indicate that the heating system will be located in the heating room in the attic. The customer's utility does not offer a special rate with blocking times. After consulting with the customer, they hand you the plans. Plan the heat pump system so that you can submit an offer.

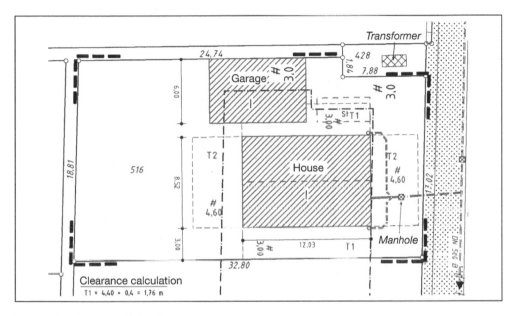

Figure 15.13 Excerpt of the site plan

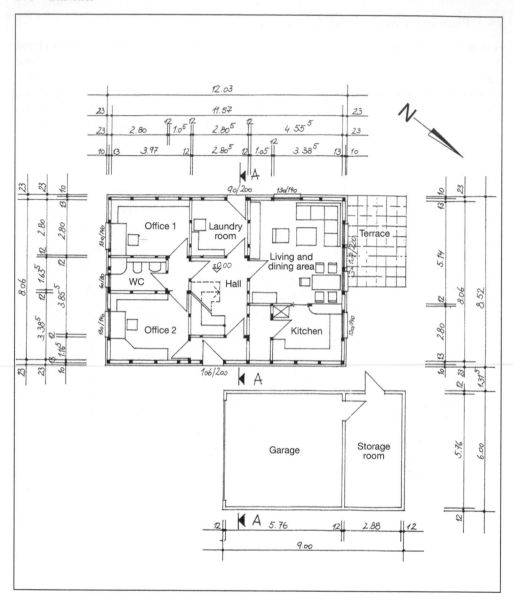

Figure 15.14 Plan of the ground floor

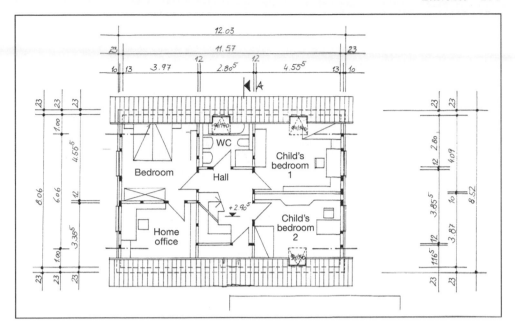

Figure 15.15 Plan of the first floor

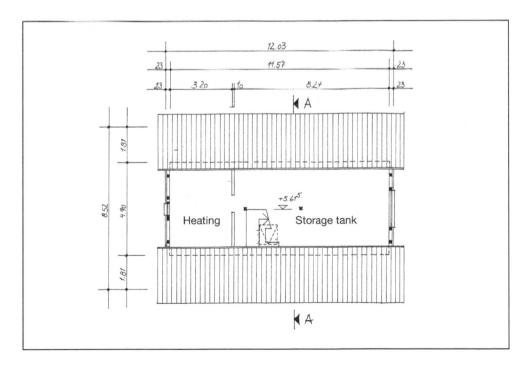

Figure 15.16 Plan of the attic

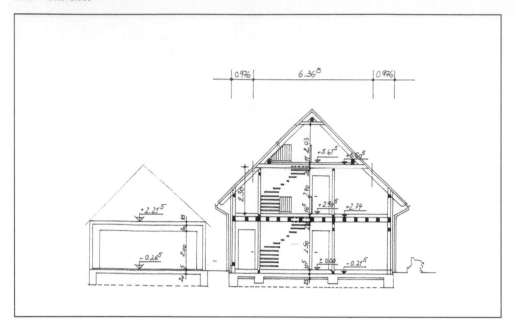

Figure 15.17 Cross-section

15.2 Planning a brine-water heat pump with borehole heat exchangers

The customer gives you a site plan and indicates the size of each of the rooms. He wants a brine-water heat pump with borehole heat exchangers and also requests cooling. The storeroom does not need to be heated. The geological subsurface is solid rock. Plan the heat pump system so that you can submit an offer.

Summary of the rooms:

Basement:

- Hobby room   37.16 m²
- Heating room   20.24 m²
- Storeroom   9.28 m²
- Corridor   7.62 m²

Ground floor:

- Living room   37.16 m²
- Kitchen   20.24 m²
- Hallway   16.06 m²
- Toilet   1.98 m²

First floor:

- First child's bedroom   17.13 m²
- Second child's bedroom   17.71 m²
- Parents' bedroom   20.24 m²
- Bathroom   9.54 m²
- Corridor   8.52 m²

- Drywall construction   56.03 m²
  in attic

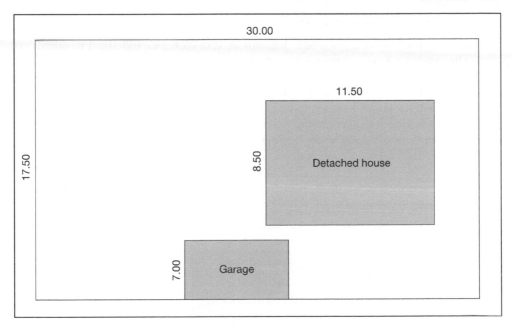

Figure 15.18 The site plan

15.3 Planning a brine-water heat pump with ground heat exchangers

A customer comes to you wanting the most cost-efficient heat pump system available. He does not want cooling. The heat pump will be installed in a prefabricated house. The temperature of the garage should be regulated because the customer has a collection of valuable classic cars; 10 °C is sufficient. The house is in the country and the plot of land is large. The customer also plans a large bath with whirlpool. The thermal insulation certificate requires a specific heat output of 40 W/m². The customer also wants the special heat pump tariff for which the power is blocked for one hour in the morning, two hours at midday, and one hour in the evening. Only two people will live in the house. There is no guest room.

Summary of the rooms:

- Heating room   8.70 m²
- Hallway   4.64 m²
- Living room   37.86 m²
- Kitchen   11.60 m²
- Bathroom   12.17 m²
- Toilet   11.67 m²
- Garage   63.76 m²

You consider an air-water heat pump, but ask the customer if he would consider undertaking the excavation work needed to install a ground heat exchanger in accordance with your instructions. The customer is happy to do so. Plan the heat pump system and draw up an offer.

15.4 Planning a larger heat pump system

You are given the following plans and diagrams by your customer and asked to submit an offer. The customer wants a heat pump system to heat his office and adjacent production hall. All office and production areas need to be cooled in summer. Now you need to advise the customer and submit an offer for the best heat pump system to suit his needs. You recommend a water-water heat pump and explain its benefits. The customer expresses a preference for a brine-water heat pump system. The groundwater level is 10 m. The ground layers to a depth of 60 m are water-bearing sand and gravel. The underlying geology at depths greater than 60 m is limestone.

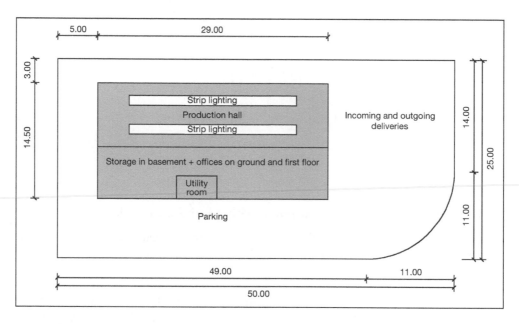

Figure 15.19 Plan

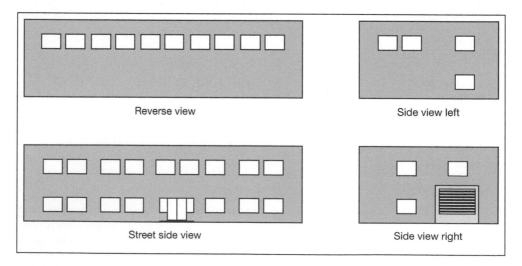

Figure 15.20 Elevations

Summary of the rooms:

Basement:

- Hallway   5.5 m^2
- Storage room 1   40.0 m^2
- Storage room 2   40.5 m^2
- Storage room 3   40.0 m^2
- Archive   10.0 m^2
- Heating room   25.0 m^2
- Corridor   22.0 m^2

Ground floor:

- Reception room   28.0 m^2
- Entry area   12.0 m^2
- Toilet – men   5.5 m^2
- Toilet – women   5.5 m^2
- Office 1   29.0 m^2
- Photocopying room   5.0 m^2
- Office 2   29.0 m^2
- Break room + shower   31.0 m^2
- Corridor   34.0 m^2

First floor:

- Toilet – men   5.5 m^2
- Toilet – women   5.5 m^2
- Manager's office   13.0 m^2
- Secretary's office   14.0 m^2
- Accounts room   8.5 m^2
- Office 3   36.5 m^2
- Office 4   29.0 m^2
- Office 5   29.0 m^2
- Corridor   36.0 m^2

Production hall:

- Floor space A$_H$ 230.5 m^2
- Hall height H$_H$ 7.5 m

Number of employees:

- Office   13
- Production hall   10

The employees working in the production hall should be able to wash and shower after their working day ends. Plan the heat pump system and draw up an offer.

15.5 Calculating the bivalence point of an air-water heat pump

Your customer needs a heat pump with a heat output of 12 kW to heat his house. Five people live in the house. There are no utility blocking times. He has already received an offer from a competitor for an air-water heat pump, model 3/11 with a heat output of 11 kW with A07/W35 – see Table 15.5. Now he wants a comparative offer from you for an air-water heat pump from the same manufacturer (see technical data). The nominal external temperature is –12 °C.

Calculate the bivalence point of the heat pump that the customer has already been offered, as well as the bivalence point of the heat pump you recommend. Show your customer the costs he can expect to pay for electricity to operate the air-water heat pump offered by your competitor, as well as for the air-water heat pump you recommend.

15.6 Calculating various energy input factors

The energy input factor is used in a building's ecological assessment. It also reflects the economic efficiency of a heat pump system.

15.6.1 Calculating the energy input factor of a detached family home with a water-water heat pump for heating the building and generating domestic hot water

The following data is known:

- Annual heating demand: Q_H = 12,000 kWh/a
- Living space to be heated: 230 m^2
- Share of heat output used for hot water generation: 20 per cent
- Underwater pump: 3.6 m^3/h
- Groundwater level including lowering: 9 m
- Charging pump output: 140 W

Please calculate the energy input factor.

15.6.2 Calculating the energy input factor of a detached family home with a brine-water heat pump for heating the building and generating domestic hot water using an electric element

The following data is known:

- Annual heating demand: Q_H = 12,000 kWh/a
- Living space to be heated: 230 m^2
- Share of heat output used for hot water generation: 20 per cent
- Three borehole heat exchangers, each 65 m, DN 25
- Groundwater level including lowering: 9 m
- Charging pump output: 140 W

Please calculate the energy input factor.

16 Questions

This section lists commonly asked questions. Refer to details provided by the various component manufacturers to answer more detailed questions.

1   How does a heat pump work?

2   Can a heat pump also be used to generate domestic hot water?

3   Is an additional form of heating required? Can a heat pump cover all the heating requirements?

4   When is a heat pump worthwhile economically? Does it make economic sense for me?

5   Do you need an electric heating element with a heat pump?

6   What are the alternatives to heat pumps?

7   What is a bivalence point?

8   What are the key factors that need to be taken into account when considering existing buildings?

9   How long will a heat pump last?

10   How high are the life cycle costs?

11   How often must a heat pump be serviced?

12   Why have a buffer tank?

13   What faults can be expected?

14   What is a low pressure fault? What causes them?

15   What is a high pressure fault? What causes them?

16   What is an evaporator?

17   What is a condenser?

18   What is a compressor?

19   What is a dryer? What is it used for?

20   What is the observation glass for?

21   What is a refrigerant collector? What is its role?

22   What is an injector?

23   What is bivalent heat pump operation?

24   What is monovalent heat pump operation?

25   What is monoenergetic heat pump operation?

26   Why should a heat pump be operated in monovalent mode?

Answer these questions and check your answers using the solutions provided in Chapter 19.

17 Examples of heat pump systems

The operator asks: Why do so few people use a heat pump?

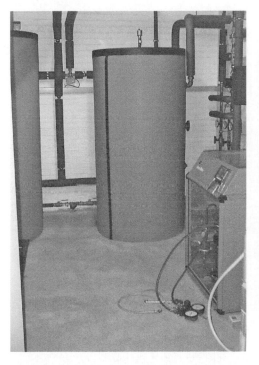

Figure 17.1
Example of a brine-water heat pump with hot water tank and buffer tank, GeoMax® model being put into operation

Figure 17.2
Example of two-stage heat pump system for high flow temperatures, manufactured by Viessmann

Figure 17.3
Example of a water-water heat pump system
with hot water tank and buffer tank, GeoMax®
model

Figure 17.4
Example of a brine-water heat pump with low-
flow solar system for hot water generation,
GeoMax® model

Figure 17.5
Two heat pumps installed according to
customer's wishes: on the left, heat pump for
heating the left side of the hall, and on the
right for heating the right half of the hall and
offices

Figure 17.6
The hall with adjacent offices

Figure 17.7
Example of a brine-water heat pump system
with hot water tank and buffer tank during
start-up in a detached house

Figure 17.8
Commissioning a heat pump in a detached
family house

17.1 A good example of local foresight in Dorsten-Wulfen

The positive example of the town of Dorsten-Wulfen with its 'cold local heating' concept
goes back to the 1970s. The plans were laid in 1977. At the time, the VEW, the regional
energy supplier in North Rhine-Westphalia, had decided to use centralised groundwater
supplies. In 1979, the local houses and apartment houses with their water-water heat
pumps were connected to the groundwater network.

At an average specific heat demand of 90.5 W/m², the overall heating demand for the
71 houses is around 1.1 MW. The average living area of the residential units is 104.6 m².

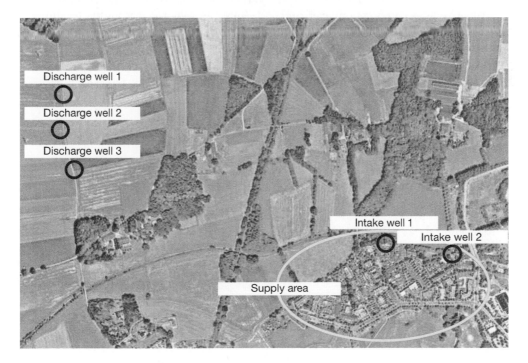

Figure 17.9 Distribution of the wells

The photo above shows the area involved and the position of the individual intake and discharge wells.

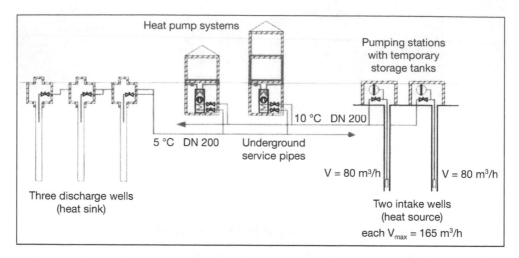

Figure 17.10 Groundwater network for decentralised heat pumps

Groundwater is drawn from two intake wells, each with a depth of around 91 m and a flow rate of around 80 m³/h. The maximum total flow rate is around 165 m³/h. The groundwater is distributed via a 1,200 m long pipe network. The houses are fed with groundwater at a constant temperature of 10 °C. Each house is equipped with underfloor heating.

Primary energy consumption levels and CO_2 emissions were the reasons for deciding on this arrangement. The primary energy consumption and CO_2 emissions were compared between heat pumps and gas and oil heating systems:

Table 17.1 Decision-making basis

	Heat pump	*Gas heating*	*Oil heating*
Net energy	100%	100%	100%
Primary energy	111%	149%	142%
Annual CO_2 emissions per m²	43 kg/m²a	54 kg/m²a	73 kg/m²a

This diagram shows that heat pumps were the most environmentally friendly solution. The following illustration also demonstrates this:

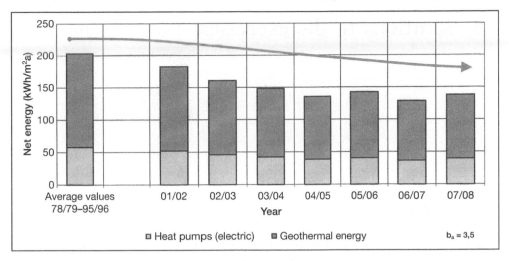

Figure 17.11 Net energy consumption

The following shows the division of costs for the heating systems:

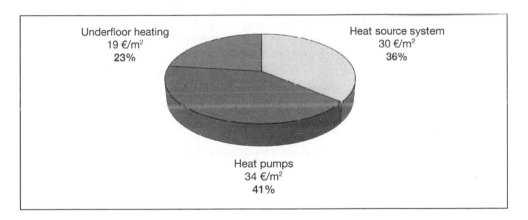

Figure 17.12 Division of costs

The German government's goal of reducing greenhouse gas emissions by 25 per cent by 2020, based on 1990 levels, requires measures to reduce all forms of energy consumption. As one-third of Germany's energy is consumed for heating buildings, generating around one-quarter of all CO_2 emissions, this represents a huge potential for savings by reducing energy demand for heating and increasing the use of renewables. The Dorsten-Wulfen concept shows that this works, even though the houses in this development do not meet today's low energy house standards. The energy supplier offers all operators the chance to use a problem-free and probably most cost-efficient heat source. The experience gained here can be used to optimise new housing developments.

18 Solutions to the exercises

IMPORTANT

All the following solutions assume an area-specific heat demand of 45 W/m². However, this is only a rough calculation, and precise calculations of heat demand must be carried out in accordance with DIN EN 12831. This may result in a lower heat output.

18.1 Planning a water-water heat pump (task 15.1)

You advise your client to use the solar system only for generating domestic hot water. The garage does not need to meet the German Energy Saving Ordinance (EnEV) requirements as it is not a living space. You agree with the customer that the garage will only be minimally heated, to low temperatures on cold days.

You discover that the boiler room is directly above the bedroom and so a very quiet heat pump is needed!

First, you need to determine the heat output. To do so, you must first calculate the total area needing heating.

Total living area requiring heating on the ground floor: 88.5 m²

TIP

Although the storeroom will not be heated, it must be included because it is located within the insulated building shell and will therefore be heated by the adjoining rooms.

The plans indicate that the garage has an upper floor. Therefore:

Total garage area needing to be heated: 79.6 m²

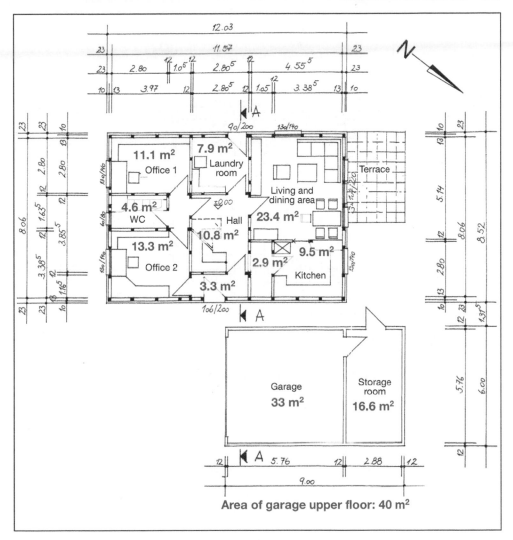

Figure 18.1 Ground floor plan

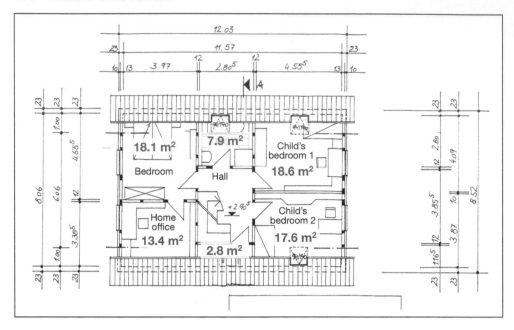

Figure 18.2 First floor plan

Total living area requiring heating on the first floor: 83.4 m².

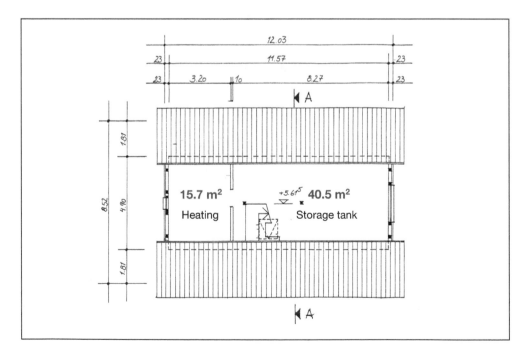

Figure 18.3 Plan of the attic

Total area to be heated in the attic: 56.2 m^2

Total area to be heated in house: 228.1 m^2

Total garage area: 79.6 m^2

The plans indicate that four people will live in the house.

How do we heat with the garage in accordance with the German Renewable Energy Heat Act (EEWärmeG)? Garages are not expressly excluded from EEWärmeG regulations. It does exclude industrial premises that need to be heated to temperatures of less than 12 °C. As the garage does not need to be warmer than 10 °C, it is permissible for it to be only slighted heated. We will take 30 W/m^2 as the area-specific heating load (although current regulations must be taken into account).

For the house with 45 W/m^2, the garage with 30 W/m^2 and the hot water generation at 300 W/person, the heat output is calculated as follows:

- 228 m^2 × 45 W/m^2 = 10.260 W   = 10.3 kW
-     + Garage 80 m^2 × 30 W/m^2 = 2.400 W   = 2.4 kW
-     + Hot water generation: 300 W/Pers. × 4 Pers. = 1.200 W = 1.2 kW

$$\overline{}$$

                                                          13.9 kW

The utility does not offer blocking times. Now you choose the heat pump from the following data sheet – (usually) one size larger than the actual required heat output.

You recommend the GeoMax® WW 14.200 heat pump with a heat output of 14.2 kW (for a flow temperature of 35 °C). You refer to the technical data to establish the minimum required flow rate on the primary (water) side: 3.4 m^3/h.

Now the wells need to be specified. Here, we use the basement floor plan. For a house of this size, the wells should be placed at a minimum distance of 15 m to 20 m from the house. As the discharge well can also be used for irrigating the garden, etc., it should be placed in an easily accessible position.

TIP

When positioning the wells, attention must be paid to the flow direction: cooler water fed into the discharge well must flow away from the intake well.

Table 18.1 Technical data for GeoMax® heat pumps

GeoMax®	WW 6.400	WW 8.300	WW 9.800	WW 11.600	WW 14.200	WW 17.500	WW 20.100	WW 23.500
Heat output								
W10/W35	6.4 kW	8.3 kW	9.8 kW	11.6 kW	14.2 kW	17.5 kW	20.1 kW	23.5 kW
W10/W55	5.9 kW	7.6 kW	9.0 kW	10.5 kW	13.0 kW	15.8 kW	18.8 kW	21.2 kW
COP								
W10/W35	5.8	5.9	5.9	5.9	5.9	5.9	6.0	6.0
W10/W55	3.2	3.2	3.2	3.3	3.4	3.4	3.4	3.4
Evaporator								
Material	Stainless steel, copper or nickel-plated brazed							
Minimum flow	1.5 m³/h	2.0 m³/h	2.3 m³/h	2.7 m³/h	3.4 m³/h	4.1 m³/h	4.7 m³/h	5.6 m³/h
Pressure loss	0.14 bar	0.23 bar	0.14 bar	0.20 bar	0.17 bar	0.26 bar	0.22 bar	0.22 bar
Connector	1"	1"	1"	1"	1"	1"		

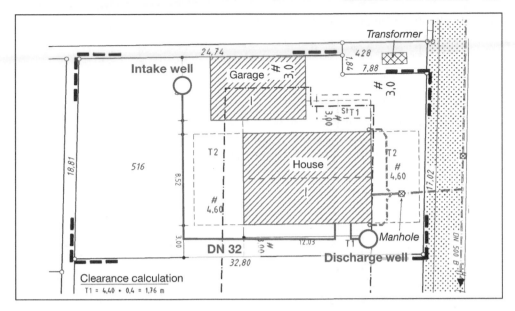

Figure 18.4 Positioning of the wells

For the piping, you select PE pipes DN 32. The overall length of both PE pipes is around 50 m, including the vertical pipes into the wells and to the heating room. According to Table 7.1, a flow rate of around 3.4 m³/h gives us a pressure loss of 15 mWs over 100 m. Therefore, for 50 m, this gives us a pressure loss of:

$$P_p = 15 \text{ mWs} / 100 \text{ m} \times 50 \text{ m}$$

$$= 7.5 \text{ mWs}$$

$$= 0.75 \text{ bar}$$

The pipes running to and from the wells will include around 10 90° bends. The 90° bends of 1″ in diameter have a pressure loss equivalent to a 1.1 m length of straight pipe. Therefore, for 10 bends, this is the equivalent of 11 m. Thus:

$$10 \text{ moulded parts} \times 1.1 \text{ m} = 11 \text{ m straight pipe}$$

$$P_{MB} = 15 \text{ mWs} / 100 \text{ m} \times 11 \text{ m} = 1.7 \text{ mWs} = 0.17 \text{ bar}$$

According to the manufacturer's specifications for heat pumps, the pressure loss over the heat pump's heat exchanger is:

$$P_{HPE} = 0.17 \text{ bar}$$

The groundwater level is 7 m below ground level. This represents a pressure loss of 7 mWs = 0.7 bar. As the water level in the well sinks once pumping starts (approx. 1–2 m), here we should use 10 mWs (= 1 bar).

This gives us an overall pressure loss of:

$$\Delta P = P_P + P_{MB} + P_{HPE} + P_H$$

$$= 0.75 \text{ bar} + 0.17 \text{ bar} + 0.17 \text{ bar} + 1 \text{ bar}$$

$$= 2.09 \text{ bar}$$

$$= 2.1 \text{ bar} = 21 \text{ mWs}$$

Now we have to choose the correct underwater pump:

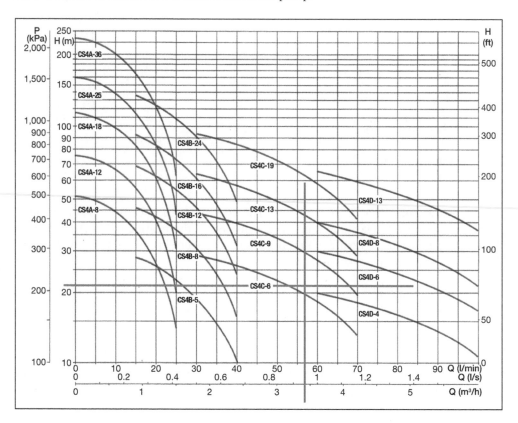

Figure 18.5 Selecting underwater pumps

As we have selected a flow rate of 3.6 m³/h rather than 3.4 m³/h for the pressure losses in the table, we now have reserves in hand. Therefore, model CS4C-6 is sufficient. Otherwise, the next size up would be recommended.

You now consult with a well engineer about the intake and discharge wells. The maximum flow rates must be taken into account. Both wells should be sized for an output of at least 4 m³/h to 5 m³/h so that an auxiliary pump can also be installed.

Now we need to plan the heat pump system on the secondary side (i.e. the heating side).

From the technical data below, you determine the minimum required water flow rate for the secondary (heating) side: 2.4 m³/h:

Table 18.2 Technical data for heat pumps

GeoMax®	WW 6.400	WW 8.300	WW 9.800	WW 11.600	WW 14.200	WW 17.500	WW 20.100
Condenser							
Material	Stainless steel, copper-plated						
Minimum flow	1.1 m³/h	1.4 m³/h	1.7 m³/h	2.0 m³/h	2.4 m³/h	3.0 m³/h	3.5 m³/h
Pressure loss	0.07 bar	0.12 bar	0.16 bar	0.10 bar	0.15 bar	0.13 bar	0.18 bar
Connector	1″	1″	1″	1″	1″	1″	1″

The pressure loss on the heating side is calculated as:

$$\Delta P_H = P_{HPC} + P_{PH} + P_{SV} + P_{MBH} + P_T$$

where P_{HPC} = pressure loss in the heat pump condenser = 0.15 bar

P_{PH} = pressure loss in the piping on the heating side

For an initial calculation, we again use the table for pressure losses.

We plan on 15 m of copper piping, DN 25:

$$P_{PH} = 7.5 \text{ mWs} / 100 \text{ m} \times 15 \text{ m} = 1.125 \text{ mWs} = 0.11 \text{ bar}$$

P_{SV} = pressure loss over the three-way switching valve, DN 25

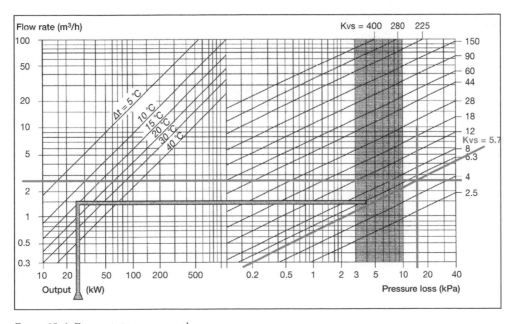

Figure 18.6 Determining pressure drop

The technical data gives us a K_{VS} value of 5.7 for the three-way switching valve DN 25. The pressure loss diagram shows a drop of 15 kPa = 0.15 bar.

Now we need to determine the pressure loss over a three-way switching ball valve with a nominal width of DN 25. Here, 2.3 m³/h = 0.64 l/s.

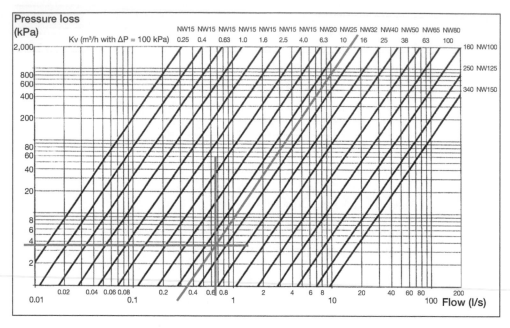

Figure 18.7 Determining pressure drop

The pressure loss over the three-way switching ball valve is almost 4 kPa = 0.04 bar. That is less than one-third of the three-way switching valve considered above.

$$P_{MBH} = \text{pressure losses over bends, moulded parts, etc.}$$

Fifteen bends and one T-shape is planned for the piping. Consequently, the pressure loss table for moulded parts gives us a pressure loss equivalent for a straight line of 15 × 1.1 m + 1 × 4.0 m = 20.5 m.

That corresponds to a pressure loss of:

$$7.5 \text{ m} / 100 \text{ m} \times 20.5 \text{ m} = 1.5 \text{ mWs} = 0.15 \text{ bar}$$

With a heat output of 14.2 kW, a high-performance hot water tank is recommended:

$$P_T = \text{pressure loss over the heat exchanger in the hot water tank, 300 l:}$$

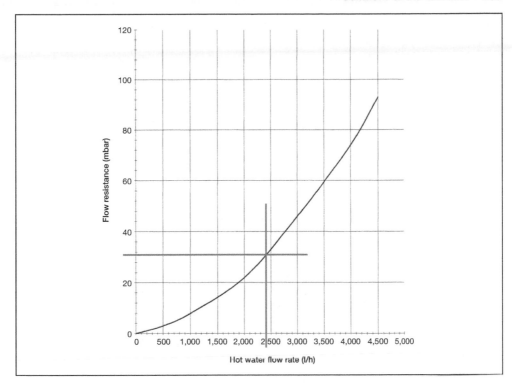

Figure 18.8 Flow resistances in the high-performance tank HRS 300

With a pressure loss of around 30 mbar = 0.03 bar

The pressure loss on the heating side with the three-way switching ball valve is calculated thus:

$$\Delta P_H = P_{HPC} + P_{LH} + P_{SV} + P_{MBH} + P_T$$
$$= 0.15 \text{ bar} + 0.11 \text{ bar} + 0.04 \text{ bar} + 0.15 \text{ bar} + 0.03 \text{ bar}$$
$$= 0.48 \text{ bar}$$
$$= 4.8 \text{ mWs}$$

We can now select the heating circulation pump using these values:

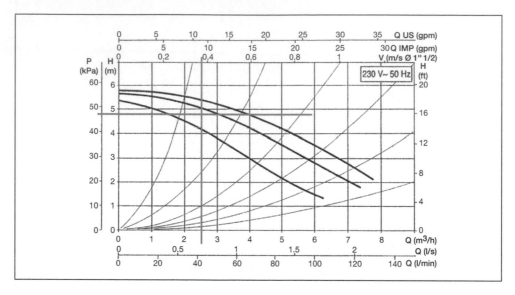

Figure 18.9 Selecting the heating circulation pump

This corresponds to the DAB heating circulation pump, model A 50/180 M. It can be operated on level 2.

TIP

If using a high-performance tank, a smaller pump can be used, especially for heat pumps with a significantly lower pressure loss.

So now we have determined the charging pump for the heating side.

The buffer tank will be sized as follows:

$$20 \text{ l/kw} \times 14.2 \text{ kW} = 284 \text{ l}$$

Thus the buffer tank should have a minimum volume of 300 l.

Now we can offer the heat pump system as follows:

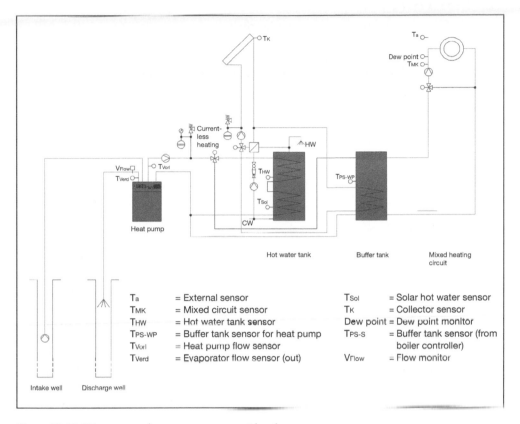

Figure 18.10 Water-water heat pump system with solar system

TIP

The new bivalent solar-hot water tanks especially designed for heat pumps (e.g. as manufactured by Austria Email AG) allow the solar system to be directly connected to the heat pump system. This removes the need for external low-flow solar heating.

A high-performance hot water tank will be installed instead of the bivalent hot water tank.

18.2 Planning a brine-water heat pump with borehole heat exchangers (task 15.2)

We must first calculate the area requiring heating:

- Basement – the customer does not wish the storeroom to be heated. However, as this room is located within the insulated building shell, we need to include it in our calculations.

- Basement: 37.16 m²
  - Utility room: 20.24 m²
  - Storeroom – only per cent, because it will not be heated directly: 4.64 m²
  - Corridor: 7.62 m²
                                                                                 69.66 m²

- Ground floor:
  - Living room: 37.16 m²
  - Kitchen: 20.24 m²
  - Hall: 16.06 m²
  - Toilet: 1.98 m²
                                                                                 75.44 m²

- First floor:
  - Child's bedroom 1:   17.13 m²
  - Child's bedroom 2:   17.71 m²
  - Parents' bedroom: 20.24 m²
  - Bathroom: 9.54 m²
  - Corridor: 8.52 m²
                                                                                 73.14 m²

- Drying room in attic: 56.03 m²
  - → Therefore, the total living space to be heated is: 274.00 m²
  - → With a specific heat output of 45 W/m², this gives us a heat output of:

    $$274 \text{ m}^2 \times 45 \text{ W/m}^2 = 12{,}330 \text{ W} = 12.3 \text{ kW}$$

    + hot water generation: 0.3 kW/Pers. × 4 Pers. = 1,200 W = 1.2 kW

    13.5 kW

There is no utility blocking period to be considered. Now you select the next largest heat pump from the price list: it is the GeoMax SW 15.000 with a heat output of 15.0 kW (for a flow temperature of 35 °C).

Table 18.3 Technical heat pump data

GeoMax®	WW 4.600	WW 5.900	WW 7.000	WW 8.300	WW 10.300	WW 12.400	WW 15.000	WW 16.800
Heat output								
B0/W35	4.6 kW	5.9 kW	7.0 kW	8.3 kW	10.3 kW	12.4 kW	15.0 kW	16.8 kW
B0/W55	4.4 kW	5.7 kW	6.7 kW	7.9 kW	9.8 kW	11.6 kW	13.9 kW	15.8 kW
Evaporator								
Material	Stainless steel, copper-plated							
Minimum flow	1.2 m³/h	1.5 m³/h	1.8 m³/h	2.0 m³/h	2.6 m³/h	3.2 m³/h	3.9 m³/h	4.4 m³/h
Pressure loss	0.10 bar	0.16 bar	0.10 bar	0.14 bar	0.20 bar	0.18 bar	0.17 bar	0.21 bar
Connector	1″	1″	1″	1″	1″	1″	1″	1″
Antifreeze	Antifrogen®							

TIP

This heat pump still has small reserves, but they would be insufficient for coping with a utility blocking time of four hours.

You then determine the minimum required brine flow rate on the primary side by referring to the technical data: 3.9 m³/h.
Now we need to plan the borehole heat exchangers. First, we need to know the cooling capacity:

$$P_C = P_{H'} \times 0.8 = 15 \text{ kW} \times 0.8 = 12 \text{ kW}$$

The geological map indicates that the local geology is solid rock with an abstraction capacity of only 35 W/m. This gives us a total borehole heat exchanger length of:

$$L = P_C / P_{Sspec} = 12{,}000 \text{ W} / 35 \text{ W/m} = 343 \text{ m}$$

After talking with the well engineer, we plan four borehole heat exchangers, each drilled to a depth of 85 or 90 m. The well engineer will use PE pipes DN 25 for the heat exchangers. Double U-pipe exchangers will be drilled. A loop has a length of 2 × 90 m = 180 m. We will need an eight-fold brine circuit manifold.

Positioning of the borehole heat exchangers:

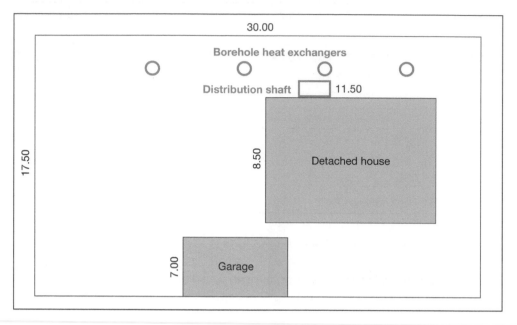

Figure 18.11 Positioning of the borehole heat exchangers

The overall pressure loss is calculated as:

$$\Delta P = P_L + P_P + P_{HEC} + P_{HPE}$$

where P_L = pressure loss over a brine loop = pressure loss over all brine loops
        P_P = pressure loss over the connecting pipes
        P_{HEC} = pressure loss over the heat exchanger cooling
        P_{HPE} = pressure loss over the heat pump heat exchanger

The pressure loss in a brine loop can be determined using the pressure loss table. First, we need to determine the brine flow rate through each brine loop:

$$Q_B = Q_{HP} / n$$

where Q_B = brine flow rate through loop
        Q_{HP} = total flow rate = minimum flow rate through the heat pump = 3.9 m³/h
        n = number of loops in the ground

For four borehole heat exchangers, this gives us eight loops through which the flow is divided. Thus:

$$Q_B = 3.9 \text{ m}^3/\text{h} / 8 = 0.49 \text{ m}^3/\text{h}$$

At 0.6 m³/h, the pressure loss is 1.8 mWs per 100 m. At 0.49 m³/h, the pressure loss will be correspondingly smaller. As the pressure loss is the square of the flow rate, the theoretical pressure loss is calculated as:

$$\Delta P_L = 1.8 \text{ mWs} / (0.6 \text{ m}^3/\text{h})^2 \times (0.49 \text{ m}^3/\text{h})^2 = 1.2 \text{ mWs}$$

For a brine loop (and therefore also for all brine loops), the pressure loss is calculated as:

$$P_L = 1.2 \text{ mWs} / 100 \text{ m} \times 180 \text{ m} = 2.16 \text{ mWs} = 0.22 \text{ bar}$$

P_P = pressure loss in the pipes running to and from the brine loops

The total length of the pipes running from the distributor to the house is 2 × 10 m = 20 m.

The feed and return pipes are DN 40.

The nominal flow rate is 3.9 m³/h. If we refer to the table for pressure loss (Table 7.1), we get a mean value of 7.5 mWs for 3.9 m³/h. Therefore, the feed and return pipes to and from the house have a maximum overall pressure loss of:

$$P_P = 7.5 \text{ mWs} / 100 \text{ m} \times 20 \text{ m}$$

$$= 1.5 \text{ mWs}$$

$$= 0.15 \text{ bar (at 3.9 m}^3/\text{h)}$$

Three 1¼" 90° bends and a distributor are needed for the connection (moulded parts). We calculate for the distributor as if it were a moulded part. The adequate length for a 90° bend is 1.1 m. For eight moulded parts, this gives us 8 × 1.2 m = 9.6 m. The pressure loss over the moulded parts is calculated as:

$$P_{MB} = 7.5 \text{ mWs} / 100 \text{ m} \times 8.8 \text{ m}$$

$$= 0.66 \text{ mWs}$$

$$= 0.07 \text{ bar (at 3.9 m}^3/\text{h)}$$

P_{HEC} = pressure loss over the heat exchanger cooling

According to the manufacturer's details, this is 20 kPa = 0.2 bar

P_{HPE} = pressure loss of the heat pump evaporator is: 0.17 bar

Thus:

$$\Delta P = P_L + P_P + P_{MB} + P_{HEC} + H_{PE}$$

$$\Delta P = 0.12 \text{ bar} + 0.15 \text{ bar} + 0.07 \text{ bar} + 0.17 \text{ bar} + 0.20 \text{ bar}$$

$$= 0.71 \text{ bar}$$

$$= 7.1 \text{ mWs}$$

This gives us a brine circulation pump manufactured by DAB, model A 80/180 M.

This brine pump must be operated on level 3.

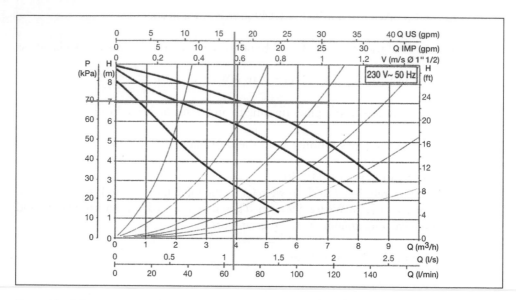

Figure 18.12 Dimensioning the brine circulation pump

Now we need to dimension the heating side:

Table 18.4 Technical data for GeoMax® heat pumps

GeoMax®	SW 4.600	SW 5.900	SW 7.000	SW 8.300	SW 10.300	SW 12.400	SW 15.000	SW 16.800
Heat output								
B0/W35	4.6 kW	5.9 kW	7.0 kW	8.3 kW	10.3 kW	12.4 kW	15.0 kW	16.8 kW
S0/W55	4.4 kW	5.7 kW	6.7 kW	7.9 kW	9.8 kW	11.6 kW	13.9 kW	15.8 kW
Condenser								
Material	Stainless steel, copper-plated							
Minimum flow	0.8 m³/h	1.0 m³/h	1.2 m³/h	1.4 m³/h	1.8 m³/h	2.1 m³/h	2.6 m³/h	2.9 m³/h
Pressure loss	0.14 bar	0.23 bar	0.09 bar	0.12 bar	0.18 bar	0.25 bar	0.17 bar	0.21 bar
Connector	1″	1″	1″	1″	1″	1″	1″	1″

The minimum flow rate on the heating side for this pump is 2.6 m³/h. Therefore, we do the same as in Section 18.1, 'Planning a water-water heat pump (task 15.1)', to calculate the charging pump on the heating side, so here is the shortened form:

The pressure loss on the heating side for piping of 1″ as in the previous calculation:

$$\Delta P_H = P_{HPC} + P_{PH} + P_{SV} + P_{MBH} + P_T$$

where P_{HPC} = pressure loss over the heat pump condenser = 0.17 bar
P_{PH} = pressure loss over the piping on the heating side, approx. 10 m, DN 25
= 9 mWs / 100 m × 10 m = 0.9 mWs = 0.09 bar
P_{SV} = pressure over the three-way switching valve, DN 25

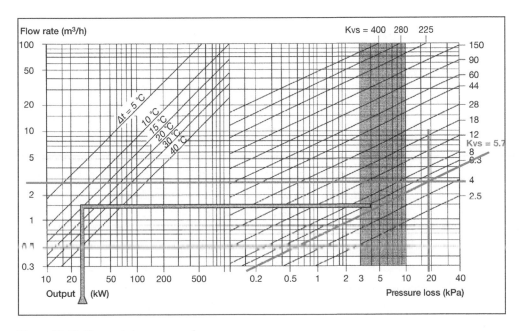

Figure 18.13 Determining pressure loss

The technical data gives us a K_{VS} value of 5.7 for the three-way switching valve, DN 25. According to the diagram for pressure loss, this gives us a pressure drop of 19 kPa = 0.19 bar.

We assume 2 × 5 bends for calculating the pressure loss over bends and T-shapes. Each bend corresponds to a length of 1.2 m. Thus we require 12 m. Therefore:

$$P_{MBH} = 9 \text{ Ws} / 100 \text{ m} \times 12 \text{ m} = 1.08 \text{ mWs} = 0.11 \text{ bar}$$

Now we need to determine pressure loss over the hot water tank. We decide on a high-performance 300 l hot water tank especially developed for heat pumps:

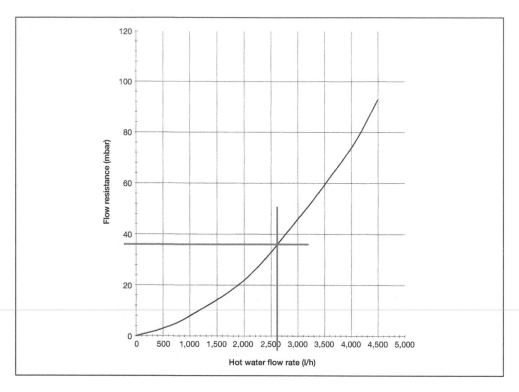

Figure 18.14 Determining pressure loss over the hot water tank

P_T = pressure loss over the hot water tank, 300 litres: 35 mbar

= 0.04 bar

The pressure loss on the heating side is calculated thus:

$\Delta P_{HWT} = P_{HPC} + P_{PH} + P_{SV} + P_{MBH} + P_T$

= 0.17 bar + 0.09 bar + 0.19 bar + 0.11 bar + 0.04 bar

= 0.60 bar

Pressure is lost over the charging pump to the buffer tank. Thus:

$\Delta P_{HBT} = P_{HPC} + P_{PH} + P_{SV} + P_{MBH}$

= 0.17 bar + 0.09 bar + 0.19 bar + 0.11 bar

= 0.56 bar

We can now determine the heating circulation pump using these values:

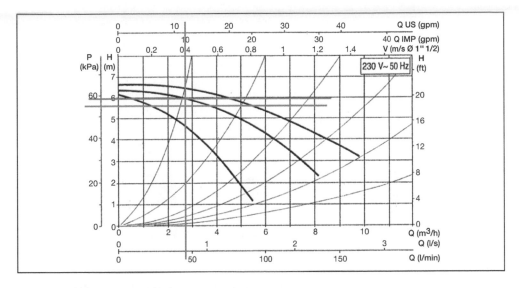

Figure 18.15 Determining the heating circulation pump

This accords with a DAB heating circulation pump, model A 56/180 M. Both pumps can be operated on level 2.

So now we have determined the charging pump on the heating side.

The buffer tank is dimensioned as follows:

20 1/kW × 15 kW = 300 1

Therefore, the buffer tank should have a minimum volume of 300 1.

Now you offer the customer the following heat pump system with cooling:

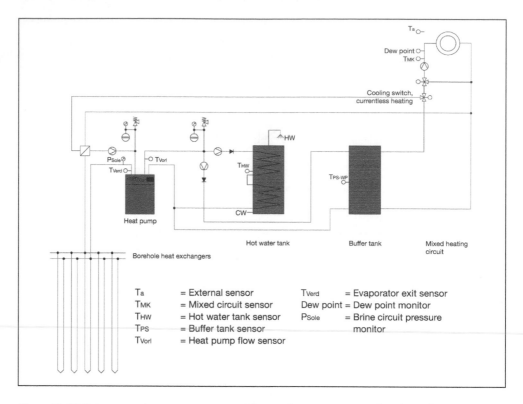

Figure 18.16 Brine-water heat pump system with two charging pumps and with cooling

Before completing planning, the electrical components must also be considered (i.e. cable cross-section, fuses, a soft starter if necessary).

18.3 Planning a brine-water heat pump with ground heat exchangers (task 15.3)

The customer wants cooling, and as the plot of land is sufficiently large you recommend ground heat exchangers.

Now you need to determine the necessary heat output:

- Heating room:
  $(0.7 \text{ m} + 1.5 \text{ m} + 0.7 \text{ m}) \times (1.15 \text{ m} + 0.1 \text{ m} + 1.75 \text{ m})$ = 8.70 m²
- Hall: $(0.45 \text{ m} + 2 \text{ m} + 0.45 \text{ m}) \times 1.6 \text{ m}$ = 4.64 m²
- Living room: 5.65 m × 6.7 m = 37.86 m²
- Kitchen: 2.9 m × 4 m = 11.60 m²
- Bathroom: $4.8 \text{ m} \times 1.75 \text{ m} + 3.25 \text{ m} \times (1.15 \text{ m} + 0.1 \text{ m})$ = 12.17 m²
- Toilet: 1.45 m × 1.15 m = 1.67 m²

_____
76.64 m²

With a specific heat output of only 35 W/m², the heat output for the house is calculated as:

$$P_H = F \times 35 \text{ W/m}^2$$
$$= 77 \text{ m}^2 \times 35 \text{ W/m}^2$$
$$= 2{,}695 \text{ W}$$
$$= 2.7 \text{ kW}$$

A large bath is planned for the bathroom and so generous quantities of hot water are needed. Therefore, you should certainly recommend a larger hot water tank (400 l). As we expect higher levels of hot water consumption, we assume a hot water requirement of 400 W/person and calculate 2 × 400 W = 800 W = 0.8 kW.

Garage: 5.9 m × 8.9 m + (3.5 m + 0.5 m + 3.5 m) × 1.5 m = 63.76 m³

If we take 25 W/m² for the garage, based on its lower required temperature, then this gives us an output of 1.6 kW.

Thus for the total heat output:

$$P_{H'} = 2.7 \text{ kW} + 0.8 \text{ kW} + 1.6 \text{ kW} = 5.1 \text{ kW}$$

Taking into account the utility blocking time of four hours, the heat pump needs a heat output of:

$$P_{H''} = P_{H'} \times 24 \text{ h} / (24 \text{ h} - t_{Off})$$
$$= 5.5 \text{ kW} \times 24 \text{ h} / (24 \text{ h} - 4 \text{ h})$$
$$= 6.1 \text{ kW}$$

As the customer is looking for a particularly inexpensive heat pump system, you might think that an air-water heat pump would be the ideal solution. However, this is not the case because a small brine-water heat pump is significantly more economically effective than an air-water heat pump. And as the customer can excavate the ground for the ground heat exchanger themselves, not only does this make installing a brine-water heat pump cheaper, but the better seasonal performance factor (SPF) means operating costs are also lower.

Table 18.5 Technical data for GeoMax® heat pumps

GeoMax®	SW 4.600	SW 5.900	SW 7.000	SW 8.300	SW 10.300	SW 12.400	SW 15.000	SW 16.800
Heat output								
BO/W35	4.6 kW	5.9 kW	7.0 kW	8.3 kW	10.3 kW	12.4 kW	15.0 kW	16.8 kW
SO/W55	4.4 kW	5.7 kW	6.7 kW	7.9 kW	9.8 kW	11.6 kW	13.9 kW	15.8 kW
Evaporator								
Material	Stainless steel, copper-plated							
Minimum flow	1.2 m³/h	1.5 m³/h	1.8 m³/h	2.0 m³/h	2.6 m³/h	3.2 m³/h	3.9 m³/h	4.4 m³/h
Pressure loss	0.10 bar	0.16 bar	0.10 bar	0.14 bar	0.20 bar	0.18 bar	0.17 bar	0.21 bar
Connector	1″	1″	1″	1″	1″	1″	1″	1″
Antifreeze	Antifrogen®							

We will need the next largest heat pump (i.e. model SW 7.000 with 7.0 kW).

The next step is to calculate the cooling capacity:

$$P_C = P_{H''} \times 0.8 = 7.0 \text{ kW} \times 0.8 = 5.6 \text{ kW}$$

As the heat pump with a heat output of 7 kW is sufficiently large, we can assume an annual operating time of 1,800 h/a for planning the ground heat exchangers.

The upper ground layers consist of a mixture of damp sand and loam. You choose the median value for the dry clay and loam mixture for the extraction capacity: 20 W/m² to 25 W/m², therefore 22.5 W/m². So for the required area, this gives us:

$$A_C = P_C / P_{Cspec}$$

$$= 5{,}600 \text{ W} / 22.5 \text{ W/m}^2$$

$$= 249 \text{ m}^2$$

$$= 250 \text{ m}^2$$

Now we have to determine the distances between the heat exchanger loops:

$$d_{L18} = 1.17 \text{ m} - P_{Cspec} \times 0.017 \text{ m/W}$$

$$= 1.17 \text{ m} - 22.5 \text{ W/m}^2 \times 0.017 \text{ m/W}$$

$$= 0.788 \text{ m}$$

$$= 0.8 \text{ m}$$

Or using the diagram:

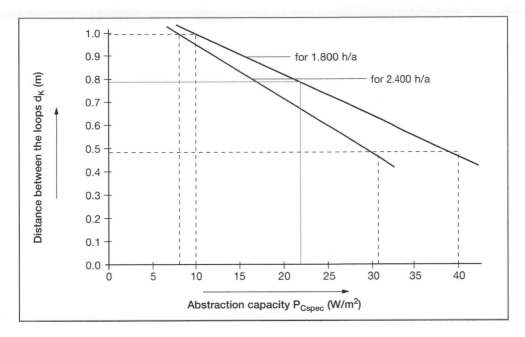

Figure 18.17 Determining distances between the loops

Now we need to plan the individual loops for the ground heat exchangers: the ground heat exchangers require a total area of at least 250 m² and so you choose an area of at least 19 m × 13 m. A length of 19 m is certainly suitable for loops of 200 m in length plus connection lengths of 2 × 5 m.

The ground heat exchangers can be laid out as follows:

1   Distribution shaft
2   Ground heat exchangers, consisting of two circuits.

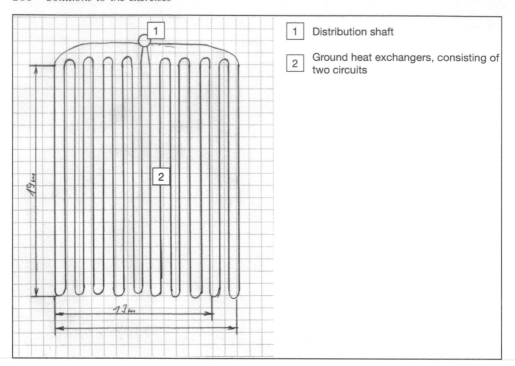

Figure 18.18 Sample ground heat exchanger layout

It is always worth making a drawing of the ground heat exchanger layout to determine whether it fits into the space available.

The connecting pipes should be as far apart from one another as possible in order to avoid excessive cooling (freezing) in a localised area.

When you make the sketch, you realise that it makes sense to choose a slightly wider exchanger width – assuming space is available. If necessary, a loop could be removed from each side. However, if the PE pipes have already been ordered and delivered, then there is no reason not to use them (apart from the minimally higher pressure loss).

In this case, we need two rolls of PE pipes, each 200 m in length. PE pipes of DN 25 are recommended for this application.

The overall pressure loss is again calculated as:

$$\Delta P = P_L + P_P + P_{MB} + P_{HPE}$$

where P_L = pressure loss over a brine loop = pressure loss over all brine loops

The pressure loss over a brine loop can be determined by referring to the pressure loss table (Table 7.1). Before doing so, we must calculate the brine flow rate per brine loop:

$$Q_B = Q_{HP} / n$$

where Q_B = brine flow rate through the loop
Q_{HP} = total flow = minimum flow rate through the heat pump
n = number of underground loops

The brine-water heat pump SW 7.000 requires a minimum flow rate of 1.8 m³/h.

$$Q_B = 1.8 \text{ m}^3/\text{h} / 2 = 0.9 \text{ m}^3/\text{h}$$

According to pressure loss Table 7.1, the pressure loss at 0.9 m³/h and 200 m is:

$$P_L = (4 \text{ mWs} / 100 \text{ m}) \times 200 \text{ m} = 0.8 \text{ bar}$$
P_P = pressure loss in the feed and return pipes

The total length of the feed and return piping from the distribution shaft to the house is 2 × 10 m = 20 m.

The feed pipes have a diameter of DN 32.
   Therefore, the feed and return piping has a total pressure loss of:

$$P_P = (4.6 \text{ mWs} / 100 \text{ m}) \times 20 \text{ m} = 0.92 \text{ Ws} = 0.09 \text{ bar}$$

Installation of the feed and return piping to the heat pump requires 2 × 3 = 6 90° bends of 1″. The necessary length is: 1.1 m × 6 = 6.6 m.

Thus:

$$P_{MB} = (4.6 \text{ mWs} / 100 \text{ m}) \times 6.6 \text{ m} = 0.3 \text{ mWs} = 0.03 \text{ bar}$$

P_{HPE} = pressure loss over the heat pump evaporator is: 0.10 bar

Thus:

$$\Delta P = P_L + P_P + P_{MB} + P_{HPE}$$

$$= 0.08 \text{ bar} + 0.09 \text{ bar} + 0.03 \text{ bar} + 0.10 \text{ bar}$$

$$= 1.02 \text{ bar}$$

$$= 10.2 \text{ mWs}$$

PLEASE NOTE

You realise that this pressure loss is rather high and so you select a PE pipe of DN 32 for the ground heat exchangers, and DN 40 for the connections, and recalculate:

$P_L = 2.28 \text{ mWs} = 0.23 \text{ bar}$

$P_P = (1.9 \text{ mWs} / 100 \text{ m}) \times 20 \text{ m} = 0.38 \text{ Ws} = 0.04 \text{ bar}$

Installation of the feed and return pipes to the heat pump requires 2 × 3 = 6 90° bends of 1″. The necessary length is: 1.1 m × 6 = 6.6 m.
   Thus:

$$P_{MB} = (1.9 \text{ mWs} / 100 \text{ m}) \times 6.6 \text{ m} = 0.13 \text{ mWs} = 0.01 \text{ bar}$$

$$P_{HPE} = \text{pressure loss over the heat pump evaporator is: } 0.10 \text{ bar}$$

Therefore, for the larger pipe cross-section:

$$\Delta P = P_L + P_P + P_{MB} + P_{HPE}$$

$$= 0.23 \text{ bar} + 0.04 \text{ bar} + 0.01 \text{ bar} + 0.10 \text{ bar}$$

$$= 0.38 \text{ bar}$$

$$= 3.8 \text{ mWs!}$$

That is significantly less, meaning fewer losses, which in turn reduces power consumption!

The brine circulation pump you need is:

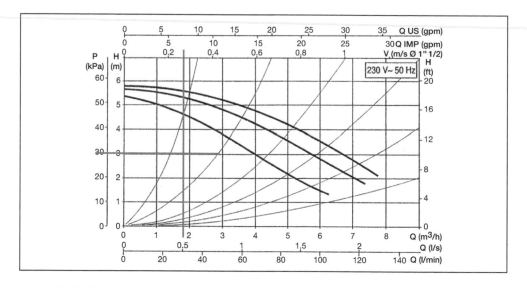

Figure 18.19 Dimensioning the brine circulation pump

This gives us the FAB brine circulation pump model A 50/180.
   This brine circulation pump must be operated on level 1. A smaller pump may also be appropriate.
   The minimum flow on the heating side for this heat pump is 1.2 m³/h (see Table 7.1).
   The pressure loss on the heating side for a pipe of ¾″ diameter is calculated with:

$$\Delta P_H = P_{HPC} + P_{LH} + P_{SV} + P_{MBH} + P_T$$

where P_{HPC} = pressure loss over the heat pump condenser = 0.09 bar
          P_{LH} = pressure loss over the piping on the heating side

For a first approximation, we again use the pressure loss table.

     10 m of ¼″ copper piping is planned.

  → P_{PH} 6.4 mWs / 100 m × 10 m = 0.64 mWs = 0.06 bar

     P_{SV} = pressure loss over the three-way switching valve, DN 20

The technical data indicates that a K_{VS} value of 5.7 applies for a three-way switching valve DN 20. Referring to the pressure loss diagram, this gives us a pressure drop of over 3.0 kPa (i.e. approx. 0.03 bar).

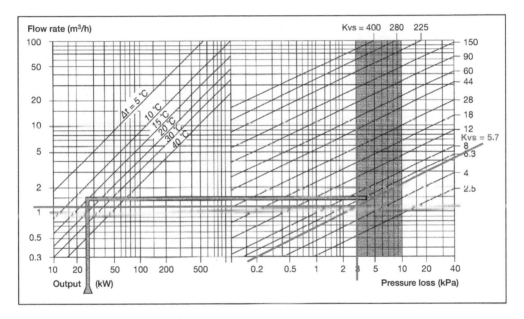

Figure 18.20 Determining pressure loss

     P_{MBH} = pressure loss over moulded parts, bends, T-shapes, etc.

Fifteen bends and one T-shape are planned for the piping. Therefore, the pressure loss table for moulded parts gives us a pipe length equivalent of:

  → P_{PH} = 20.5 mWs / 100 m × 10 m = 2.05 mWs = 0.21 bar

     P_T = pressure loss over the heat exchanger in the hot water tank, 300l:

Taking a pressure loss of around 0.06 bar + 0.02 bar at 1.1 m³/h.

Table 18.6 Pressure loss over the hot water tank

	Heating area (m²)	N_L-number (Thermostat setting 60°C)	70°C / 45°C / 10°C — 1 m³/h	2 m³/h	3 m³/h	80°C / 45°C / 10°C — 1 m³/h	2 m³/h	3 m³/h	70°C / 60°C / 10°C — 1 m³/h	2 m³/h	3 m³/h	80°C / 60°C / 10°C — 1 m³/h	2 m³/h	3 m³/h	Flow resistance in mbar — 1 m³/h	2 m³/h	3 m³/h
			Throughput in kW or l/h														
BE 160 ERM	0,60	2	11,9 / 293	13,9 / 342	14,9 / 367	15,6 / 384	18,5 / 455	19,8 / 487	8,5 / 146	9,7 / 167	10,3 / 177	12,5 / 215	14,7 / 253	15,7 / 270	2	14	54
BE 200 ERM	1,00	3,5	18,0 / 443	21,7 / 534	23,5 / 578	23,3 / 573	28,4 / 699	31,0 / 763	13,2 / 227	15,5 / 267	16,6 / 286	19,1 / 329	22,9 / 394	24,8 / 427	14	45	96
BE 300 ERM	1,50	7,5	23,0 / 566	30,1 / 740	31,8 / 782	29,8 / 733	39,1 / 962	42,7 / 1050	17,1 / 294	20,9 / 360	22,4 / 386	24,8 / 427	31,0 / 534	33,9 / 584	32	90	178
BE 400 ERM	1,80	11	27,2 / 669	34,8 / 856	38,9 / 957	35,1 / 863	45,1 / 1109	50,7 / 1247	20,4 / 351	25,5 / 439	27,5 / 474	29,3 / 505	37,1 / 639	41,2 / 709	53	114	210
BE 500 ERM	2,00	15	29,8 / 733	39,2 / 964	44,2 / 1087	38,3 / 942	51,2 / 1260	58,1 / 1429	21,9 / 377	27,2 / 468	29,5 / 508	31,7 / 546	42,1 / 725	48,1 / 826	41	139	293
BE 300 ERMR below	1,50	7,5	23,0 / 566	30,1 / 740	31,8 / 782	29,8 / 733	39,1 / 962	42,7 / 1050	17,1 / 294	20,9 / 360	22,4 / 386	24,8 / 427	31,0 / 534	33,9 / 584	32	90	178
BE 300 ERMR above	1,00	1,8	16,6 / 408	20,2 / 497	21,8 / 536	21,9 / 539	26,7 / 657	29,1 / 716	12,2 / 210	14,4 / 248	15,7 / 270	18,1 / 312	21,7 / 374	23,6 / 406	20	58	121
BE 400 ERMR below	1,80	11	27,2 / 669	34,8 / 856	38,9 / 957	35,1 / 863	45,1 / 1109	50,7 / 1247	20,4 / 351	25,5 / 439	27,5 / 474	29,3 / 505	37,1 / 639	41,2 / 709	53	114	210
BE 400 ERMR above	1,00	3	16,7 / 411	20,0 / 492	21,5 / 529	21,6 / 531	26,1 / 642	28,2 / 694	12,4 / 214	14,5 / 250	15,4 / 265	18,0 / 310	21,4 / 369	23,0 / 396	12	40	83
BE 500 ERMR below	2,00	15	29,8 / 733	39,2 / 964	44,2 / 1087	38,3 / 942	51,2 / 1260	58,1 / 1429	21,9 / 377	27,2 / 468	29,5 / 508	31,7 / 546	42,1 / 725	48,1 / 828	41	139	293
BE 500 ERMR above	1,00	3,7	16,2 / 399	19,6 / 482	20,9 / 514	20,3 / 499	25,0 / 615	27,5 / 677	11,4 / 196	13,5 / 232	14,0 / 241	16,8 / 289	19,9 / 343	21,0 / 362	19	55	109

Thus: $P_T = 0.08$ bar

The pressure loss on the heating side is then calculated as:

$$\Delta P_H = P_{HPC} + P_{PH} + P_{SV} + P_{MBH} + P_T$$

$$= 0.09 \text{ bar} + 0.06 \text{ bar} + 0.03 \text{ bar} + 0.21 \text{ bar} + 0.08 \text{ bar}$$

$$= 0.55 \text{ bar}$$

Now these values can be used to select the heating circulation pump:

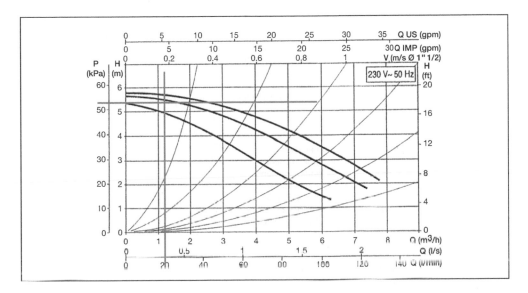

Figure 18.21 Determining the brine circulation pump

This gives us the FAB brine circulation pump model A 50/180 M.

The pump can be operated on level 2.

TIP

Here again, just a few measures will help reduce this pressure loss significantly:

* slightly larger pipe cross-section, which will have an impact, especially for the moulded parts; and
* either a switching ball valve, or two charging pumps, although here the pressure loss over the non-return valve should be as low as possible.

The buffer tank is dimensioned as follows: 20 l/kW × 7 kW = 140 l.

Consequently, the buffer tank should have a minimum volume of 200 l.

Now you offer the customer the following heat pump system with ground heat exchangers:

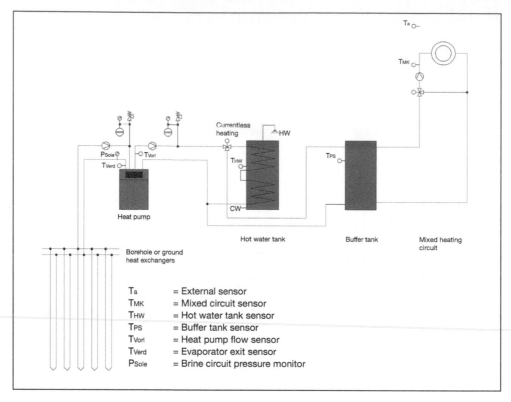

Figure 18.22 Brine-water heat pump system with cooling

Before completing planning, the electrical components must also be considered (i.e. cable cross-section, fuses, a soft starter if necessary).

18.4 Planning a larger heat pump system (task 15.4)

The customer wants a heat pump system to heat living and office space, as well as his workshop.

Now we need to determine the required heat output for the following rooms:

Basement

- Corridor   5.5 m^2
- Storeroom 1   40.0 m^2
- Storeroom 2   40.5 m^2
- Storeroom 3   40.0 m^2
- Filing room   10.0 m^2
- Utility room   25.0 m^2
- Hallway   22.0 m^2

     Total: 183.0 m^2

Ground floor

- Reception room   28.0 m^2
- Entrance hall   12.0 m^2
- Toilets – men   5.5 m^2
- Toilets – women   5.5 m^2
- Office 1   29.0 m^2
- Photocopying room   5.0 m^2
- Office 2   29.0 m^2
- Break room + shower   31.0 m^2
- Hallway   34.0 m^2

     Total: 179.0 m^2

First floor

- Toilets – men   5.5 m^2
- Toilets – women   5.5 m^2
- MD's office   13.0 m^2
- Secretarial room   14.0 m^2
- Accounts room   8.5 m^2
- Office 3   36.5 m^2
- Office 4   29.0 m^2
- Office 5   29.0 m^2
- Hallway   36.0 m^2

Total: 177.0 m^2

Production hall

- Floor area   230.0 m^2
- Hall height   7.5 m

Strip lighting is planned for the production hall ceiling. You recommend that the ceiling contains openings so that hot air can escape during the summer. For winter, you recommend propellers or other ventilation equipment to force the hot air gathering under the ceiling back down into the hall. This helps distribute the heat through the hall and reduces heat loss beneath the ceiling.

The entire area to be heated – excluding the hall – amounts to: 539.0 m^2

This gives us a heat output as follows:

$$539 \text{ m}^2 \times 45 \text{ W/m}^2 = 24.3 \text{ kW}$$

+ hot water generation for toilets, offices + social areas, showers for 10 people as follows:

$$200 \text{ W/person} \times 10 \text{ people} = 2.0 \text{ kW}$$

→   Total heat output: 26.3 kW

TIP

We take 200 W/person because hot water consumption in commercial objects is lower than in residential homes.

Now we calculate the heat output needed for the production hall. The hall temperature is not permitted to fall below 18 °C under external temperatures of –12 °C (see Section 5.2.4, 'Determining heat pump output for industrial halls').

$$Q = V_H \times q \times \Delta T \times f_1 \times f_2 \times f_3 \times f_4$$

First, we need to calculate the hall volume:

$$V_H = L_H \times W_H \times H_H$$
$$F_H = L_H \times W_H$$

$$V_H = F_H \times H_H = 230 \text{ m}^2 \times 7.5 \text{ m}$$
$$= \text{approx. } 1{,}725 \text{ m}^3$$

The following diagram is used to determine the specific heat demand:

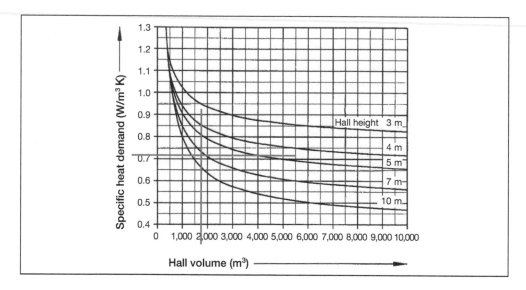

Figure 18.23 Determining specific heat demand

$\rightarrow$ $q = 0.72$ W/m³K

The side view shows that the window and doors take up over 10 per cent of the surface area. As the hall is a new build, we can assume insulation in accordance with the German 2009 Energy Saving Ordinance (EnEV). Taking into account the large proportion of surface area taken up by windows and doors, we take:

$\rightarrow$ $f_1 = 0.7$

The hall is built in a normal environment. As it is a production hall, we can assume doors will occasionally be open for up to 5 min/h. No door coverings or airwall systems are planned.

→ $f_2 = 1.4$

The hall only has one longitudinal wall adjoining a heated area.

→ $f_3 = 0.85$

This is a hall of standard proportions with an aspect ratio of 1:4 to 2:3.

→ $f_4 = 1.0$

The heat demand for the hall is calculated as:

→ $Q = V_H \times q \times \Delta T \times f_1 \times f_2 \times f_3 \times f_4$

$\quad = 1,725\ m^3 \times 0.72\ W/m^3K \times 32\ K \times 0.7 \times 1.4 \times 0.85 \times 1.0$

$\quad = 33.1\ kW$

This gives us an overall heat demand of: 26.3 kW + 33.1 kW = 59.4 kW

The utility offers the customer a special tariff that includes three blocking times distributed throughout the day and amounting to a total of 4 hours (2 × 1 + 1 h × 2 h). This must also be taken into account.

→ required heat pump output: 59.4 kW × 1.2 = 71.3 kW

You recommend two water-water heat pumps, each with a heat output of 36.5 kW:

Table 18.7 Technical data for heat pumps

GeoMax®	SW 26.300	SW 31.800	SW 36.500	SW 45.200	SW 55.100
Heat output					
W10/W35	26.3 kW	31.8 kW	36.5 kW	45.2 kW	55.1 kW
W10/W55	24.1 kW	29.0 kW	33.5 kW	41.2 kW	50.1 kW
Evaporator					
Material	Stainless steel, copper or nickel-plated brazed				
Minimum flow	6.3 m³/h	7.6 m³/h	8.7 m³/h	10.7 m³/h	13.0 m³/h
Pressure loss	0.28 bar	0.31 bar	0.32 bar	0.34 bar	0.35 bar
Connector	1¼"	1¼"	1¼"	1½"	1½"
Condenser					
Material	Stainless steel, copper plated				
Minimum flow	4.5 m³/h	5.5 m³/h	6.3 m³/h	7.8 m³/h	9.5 m³/h
Pressure loss	0.20 bar	0.20 bar	0.21 bar	0.25 bar	0.28 bar
Connector	1¼"	1¼"	1¼"	1½"	1½"

Now we need to position the intake and discharge well. For this level of output, they should be separated by a minimum of 35 m to 40 m. The intake well must be fitted with two underwater pumps.

The distance, positioning and installation of both wells depends on the permeability of the ground and the direction of flow.

The plot of land is large enough. You get a well engineer to quote for well installation and together you select the position of the wells:

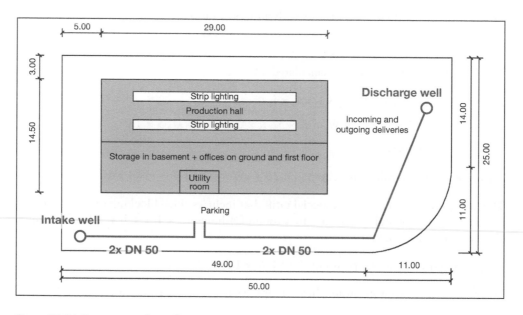

Figure 18.24 Positioning the wells

Groundwater level: 10 m

Let us look at the heat pumps individually:

The overall length of the PE pipes, including piping in the wells, is approx. 14 m + 17 m + 4 m + 3 m + 3 m + 4 m + 23 m + 22 m + 12 m = 102 m.

With a joint flow rate of 8.7 m³/h, the pressure loss table for PE piping DN 50 gives us a pressure loss of maximum 6 m per 100 m (i.e. 0.6 bar over 100 m).

→   PP = 8.6 mWs at 100 m = 0.86 bar

A total of 10 90° bends are planned for the feed and return piping. The 90° bends of 2″ correspond to a pressure loss over a straight pipe of 1.4 m. For 10 bends, this is 14 m. The pressure loss table again gives us a pressure loss of 6 mWs over 100 m.

→   10 moulded parts × 1.4 m = 14 m straight piping

→ P_{MB} = 8.6 mWs / 100 m × 14 m = 1.20 mWs = 0.12 bar

For pressure loss over the heat pump heat exchanger:

$$P_{HPE} = 0.32 \text{ bar}$$

The groundwater level is 10 m. This corresponds to a height drop of 10 mWs = 1 bar.

If the return piping is fed into the discharge well below the groundwater level, then this value can be ignored (principle of communicating pipes). As soon as the underwater pump is turned on, the groundwater level in the well will fall. There will also be natural fluctuations in well water height.

To be on the safe side, we will assume a fall in water level of 5 m (which would be a lot). Therefore, we assume an additional loss of 1 bar + 0.5 bar = 1.5 bar.

For the cooling:

$$P_{HEC} = 20 \text{ kPa} = 0.2 \text{ bar}$$

This gives us a total pressure loss of:

$$\Delta P = P_P + P_{MB} + P_{HP} + P_H + P_{HEC}$$
$$= 0.86 \text{ bar} + 0.12 \text{ bar} + 0.32 \text{ bar} + 1.5 \text{ bar} + 0.2 \text{ bar}$$
$$= 3.0 \text{ bar} = 30 \text{ mWs}$$

We then select the appropriate pump:

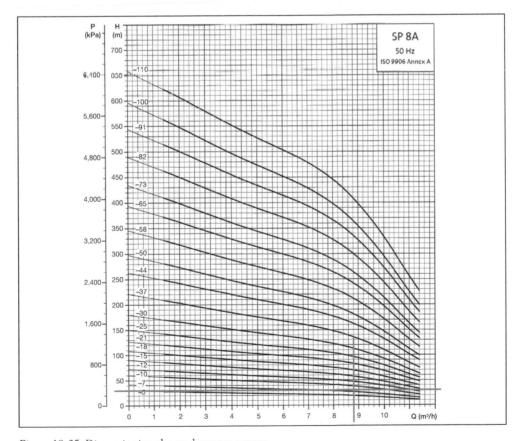

Figure 18.25 Dimensioning the underwater pumps

The underwater pump SP 8A-10 is the most suitable size.

If the flow rate is a little higher, this will have no impact on the function.

Now we need to dimension the charging pump. The nominal flow rate per heat pump is 6.3 m³/h.

Here, we can consider whether only one of the heat pumps should generate hot water, or whether they should both be used alternately. The first option requires simpler installation (= lower costs) but will not be so easy to regulate. The second option means that the heat pumps will share the load equally and redundancy is higher, which makes more sense in operating terms. Therefore, we will assume that option two is chosen (i.e. both heat pumps are feeding both storage systems).

We recommend a storage tank system for hot water generation, as the minimum flow rate on the heating side is too large for a conventional hot water tank.

Next, we consider the heat pump for heating and hot water generation.

Here again:

$$\Delta P_H = P_{HPC} + P_{PH} + P_{SV} + P_{MBH} + P_T$$

where P_{HPC} = pressure loss over the heat pump condenser = 0.21 bar
         P_{PH} = pressure loss over the piping on the heating side
         Q = 6.3 m³/h

In a first approximation, we use the pressure drop table for PE pipes.

We plan to use 15 m of 1½″ copper piping (corresponds to PE DN 40).

→ P_{PH} = 16 mWs / 100 m × 24 m

= 2.4 mWs

= 0.24 bar

P_{SV} = you select a three-way switching mixer valve, DN 32, with a K_{VS} value of 28 for this large flow volume of 6.3 m³/h.

The flow rate is 6.3 m³/h = 1.75 l/s.

Using the following diagram, we determine the pressure loss = 15 kPa = 0.15 bar

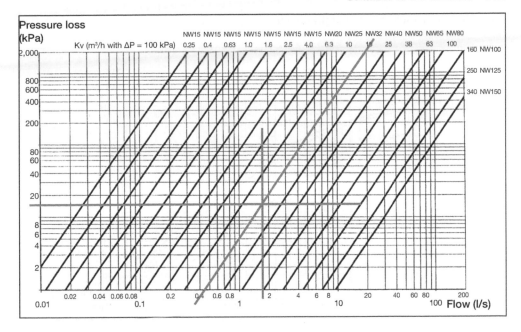

Figure 18.26 Pressure loss over the switching valve

P_{MBH} = pressure loss over bends, T-shapes, etc., 1½"

Ten bends and one T-shape are planned for the piping.
The pressure loss table for moulded parts gives us a corresponding straight length of 10 × 1.3 m + 1 × 5.0 m = 18 m.
This corresponds to a pressure loss of 16 mWs / 100 m × 18 mWs = 0.29 bar

You select a storage system for hot water generation. That means the heat pump uses a heat exchanger to charge the hot water tank. This heat exchanger is sufficiently dimensioned so that it has a pressure loss of around 0.15 bar:

$P_T = P_{CHE}$

= pressure loss over the charging heat exchanger on the primary side

= 0.15 bar

We should also consider further pressure losses for non-return valves. We make the following assumption:

P_{NRV} = 0.1 bar

This gives us a total pressure loss of:

$\Delta P_H = P_{HPC} + P_{PH} + P_{SV} + P_{MBH} + P_{CHE} + P_{NRV}$

= 0.21 bar + 0.24 bar + 0.15 bar + 0.29 bar + 0.15 bar + 0.1 bar

= 1.14 bar

This allows us to select the charging pump:

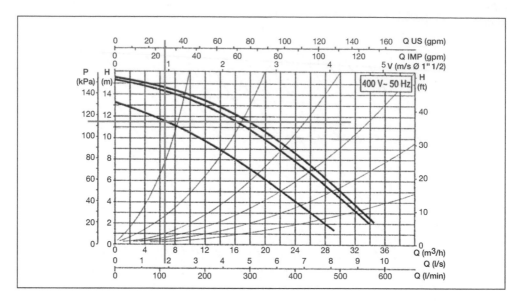

Figure 18.27 Selecting the heating circulation pump

This gives us the DAB charging pump model BPH 150/250.50 T.

This charging pump for heating and hot water generation can be operated on level 1. The optimal operating point needs to be determined by measuring the temperatures when starting up the system.

For larger heat pump systems, it makes sense for the hot water and buffer tanks to each have their own charging pump, rather than using a single charging pump and a switching valve. It is also recommended that the heat pumps are each assigned their own charging pump for the hot water and buffer tank, because usually only one of the two heat pumps runs during the transition periods. Then only a small circulation pump is required, which again saves energy.

In this case, there would be no pressure loss over a switching valve (i.e. $P_{SV} = 0$).

We continue to consider pressure losses over the larger piping sizes (i.e. 1¼"):

$\rightarrow$ P_{PH} = 4.6 mWs / 100 m × 24 m = 2.4 mWs = 1.104 mWs = 0.11 bar

and P_{MBH} = 4.6 mWs / 100 m × 18 mWs = 0.828 mWs = 0.08 bar

$\rightarrow$ ΔP_{HHW} = P_{HPC} + P_{PH} + P_{SV} + P_{MBH} + P_{CHE} + P_{NRV}

= 0.21 bar + 0.11 bar + 0 bar + 0.08 bar + 0.15 bar + 0.1 bar

= 0.65 bar

Now we can select the charging pump:

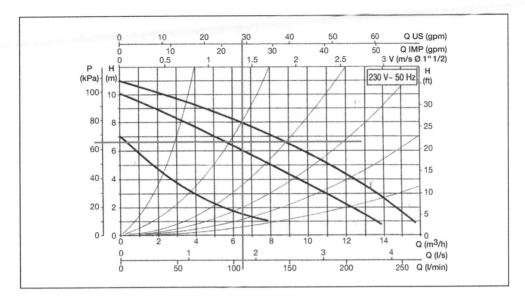

Figure 18.28 Selecting the heating circulation pump

This gives us the DAB charging pump model BPH 120/250.40 M.

This charging pump for heating and hot water generation must be operated on level 3. The optimal operating point must be determined during start-up by measuring the temperatures. If necessary, the pump can also be operated at level 2.

• These observations demonstrate that, with a little thought, the hydraulic losses can be significantly reduced, here by making the piping a little larger and removing the need for a switching ball valve. It is always worthwhile considering the possibilities.

So this smaller charging pump is sufficient, which means reduced electricity consumption for the operator.

It might also be worth considering offering the customer two high-performance 300 l hot water tanks charged in parallel. This means more hot water is on hand and there is no need for the heat exchanger and second charging pump on the secondary side. That also helps save energy.

Therefore, for charging the buffer tank, with 1¾" piping and no switching valve:

$$\Delta P_{HBT} = P_{HPC} + P_{PH} + P_{SV} + P_{MBH} + P_{BT} + P_{NRV}$$

$$= 0.21 \text{ bar} + 0.11 \text{ bar} + 0 \text{ bar} + 0.08 \text{ bar} + 0 \text{ bar} + 0.1 \text{ bar}$$

$$= 0.5 \text{ bar}$$

Now we can determine the charging pump:

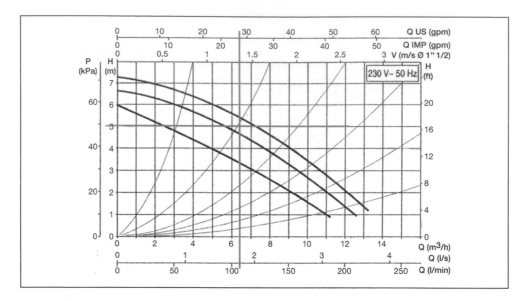

Figure 18.29 Selecting the heating circulation pump

This gives us DAB charging pump model BPH 60/250.40 M.

This charging pump for heating and hot water generation should be operated on level 3, although it can also be operated on level 2. The optimal operating point must be determined during start-up by measuring the temperatures.

Now we need to consider and dimension the domestic water pump for charging the tank.

The secondary pressure loss over the charging heat exchanger is significantly lower because the flow rate is reduced by around one quarter:

$$P_{HECs} = 0.05 \text{ bar}$$

We select a flow rate of QST = 6.3 m³/h / 4 = 1.6 m³/h for charging the tank.

As the charging heat exchanger is installed directly next to the storage tank, then the pressure losses over the pipes can almost be ignored. A pressure loss of 0.01 bar will suffice. Therefore, we calculate the overall pressure loss on the primary side as:

$$\Delta P_{STC} = 0.05 \text{ bar} + 0.01 \text{ bar} = 0.06 \text{ bar}$$

Thus a small pump is sufficient for the charging pump on the secondary side:

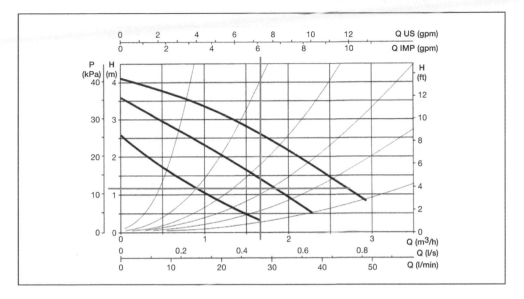

Figure 18.30 Dimensioning the storage tank charging pump

This gives us a DAB charging pump model VA 55/130·1/2.

This circulation pump is operated on level 2.

IMPORTANT

As this pump is transporting drinking water, it must be approved for this purpose (i.e. it needs to be a bronze pump)!

The buffer tank is dimensioned as follows:

$$20 \text{ l/kW} \times 2 \times 36.5 \text{ kW} = 1{,}460 \text{ l}$$

The planning here is only for the output of one heat pump. During the transition period, only one of the two heat pumps would be working. They would work alternately. Only when temperatures are low would both heat pumps operate simultaneously, with a correspondingly large quantity of heat being extracted so that the buffer tank takes on a hydraulic function.

We recommend a buffer tank with a minimum capacity of 1,000 l.

The following diagram shows the heat pump system being discussed – water-water heat pump system with charging system and cooling:

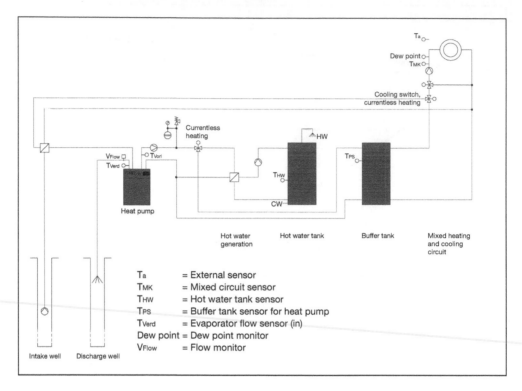

Ta = External sensor
TMK = Mixed circuit sensor
THW = Hot water tank sensor
TPS = Buffer tank sensor for heat pump
TVerd = Evaporator flow sensor (in)
Dew point = Dew point monitor
VFlow = Flow monitor

Figure 18.31 Water-water heat pump system with switching valve and charging system and cooling

We must take into account that two heat pumps will be operating here, rather than one.

The following shows a heat pump system with charging pumps rather than switching values:

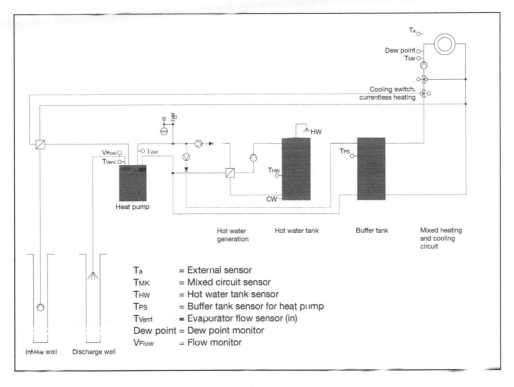

Figure 18.32 Water-water heat pump system with two charging pumps and charging system and cooling

We must take into account that two heat pumps will be operating here, rather than one.

We are not showing the heat pump system with two heat pumps and two high-performance hot water tanks being operated in parallel.

TIP

As already noted, it makes sense to assign with one charging pump for hot water generation and one for the buffer tank to each heat pump. This means that when only one heat pump is operating, only one of the two smaller pumps needs to run, rather than a single large one. This saves energy.

Now we have to offer the customer a brine-water heat pump with borehole heat exchangers as an alternative. The required heat output as calculated above, and taking the utility blocking times into account, is again:

$$P_{H''} = 71.3 \text{ kW}$$

Now we refer to the technical data sheets to choose the heat pumps:

Table 18.8 Technical data for brine-water heat pumps

GeoMax®	SW 18.800	SW 23.100	SW 26.800	SW 33.100	SW 40.100
Heat output					
B0/W35	18.8 kW	23.1 kW	26.8 kW	33.1 kW	40.1 kW
B0/W55	17.9 kW	21.9 kW	25.1 kW	31.1 kW	38.0 kW
Evaporator					
Material	Stainless steel, copper-plated				
Minimum flow	4.8 m³/h	6.0 m³/h	7.0 m³/h	8.6 m³/h	10.4 m³/h
Pressure loss	0.19 bar	0.21 bar	0.23 bar	0.28 bar	0.27 bar
Connector	1¼"	1¼"	1¼"	1½"	1½"
Antifreeze	Antifrogen® – 30%				
Condenser					
Material	Stainless steel, copper plated				
Minimum flow	3.2 m³/h	4.0 m³/h	4.6 m³/h	5.7 m³/h	6.9 m³/h
Pressure loss	0.26 bar	0.23 bar	0.30 bar	0.30 bar	0.32 bar
Connector	1¼"	1¼"	1¼"	1½"	1½"

We select two heat pumps, each with a heat output of 40.1 kW. Together, their output is sufficient. Theoretically, we could also choose one heat pump with 33.1 kW and the other with 40.1 kW. However, to keep things simple, we decide on two heat pumps, each 40.1 kW.

The cooling capacity needed to plan the borehole heat exchangers is calculated as:

$$P_C = P_{H''} \times 0.8 = 2 \times 40.1 \text{ kW} \times 0.8 = 2 \times 32.1 \text{ kW} = 64.2 \text{ kW}$$

Now we need to determine the number of borehole heat exchangers. We assume an average abstraction capacity of 55 W/m. This gives us an overall borehole heat exchanger length for each heat pump of:

$$L_C = P_C / P_{Cspec} = 62,200 \text{ W} / 55 \text{ W/m} = 1,130 \text{ m}$$

After discussion with the well engineer, we assume a drilling depth of 100 m. To calculate the number of borehole heat exchangers:

$$n_C = 1{,}130 \text{ m} / 100 \text{ m} = 11.3 = 12 \text{ borehole heat exchangers}$$

Therefore, we need a total of 12 borehole heat exchangers with a depth of 95 m to 100 m (i.e. six borehole heat exchangers per heat pump).

For the remaining borehole heat exchangers, we assume an abstraction capacity of 15 W/m for the upper 10 m, and 60 W/m for the water-bearing layers of sand and gravel. The abstraction capacity for each heat pump for the upper 10 m is:

$$P_{C1} = 12 \times 10 \text{ m} \times 15 \text{ W/m} = 1{,}800 \text{ W}$$

The remaining required abstraction capacity for the sand and gravel layers is calculated as:

$$P_{C2} = P_C - P_{C1} = 64{,}200 \text{ W} - 1{,}800 \text{ W} = 62{,}400 \text{ W}$$

Now we calculate the total borehole heat exchanger length for the smaller heat pump in the sand and gravel layers:

$$L_2 = P_{C2} / P_{Cspec2} = 62{,}400 \text{ W} / 60 \text{ W/m} = 1{,}040 \text{ m}$$

For n = 12 borehole heat exchangers, this gives us a borehole heat exchanger length in the sand and gravel layers of:

$$L_{BHE2} = 1{,}040 \text{ m} / 12 = 86.7 \text{ m}$$

And with L_{BHE1} = 10 m, we have an overall borehole heat exchanger length of:

$$L_{BHE} = L_{BHE1} + L_{BHE2} = 10 \text{ m} + 87 \text{ m} = 97 \text{ m}$$

All 12 borehole heat exchangers will be drilled as double U-pipe borehole heat exchangers with a nominal width of DN 25, each bored to a depth of 100 m.

Two 2 × 12-fold brine circuit manifolds are required for the two heat pumps.

Now we need to select the brine circulation pumps.

The overall pressure loss is calculated as:

$$\Delta P = P_L + P_P + P_{MB} + P_{HPE}$$

where P_L = pressure loss over a brine loop = pressure loss over all brine loops

Again, the pressure loss over a brine loop can be determined using the pressure loss table. But first, we need to determine the brine flow rate per brine loop:

$$Q_B = Q_{HP} / n$$

where   Q = brine flow rate per loop
Q_{HP} = total flow rate = minimum flow rate through both heat pumps
= 2 × 10.4 = 20.8 m³/h
n = number of loops

We select the circulation pump SW 40.100 for the brine-water heat pump.

This heat pump requires six double U-pipe borehole heat exchangers, DN 25. Therefore:

$$Q_B = 2 \times 10.4 \text{ m}^3/\text{h} / (2 \times 6) = 1.73 \text{ m}^3/\text{h}$$

The pipes running from the borehole heat exchangers to the brine circuit manifold have a length of 30 m. Using the pressure loss table, this gives us a pressure loss per loop of around:

$$P_L = 1.77 \text{ mWs} / 100 \text{ m} \times (200 \text{ m} + 2 \times 30 \text{ m}) = 4.6 \text{ mWs} = 0.46 \text{ bar}$$

$$P_P = \text{pressure loss in the feed and return pipes}$$

The overall length of the feed and return pipes from the brine circuit manifold to the house is 2×20 m = 40 m. The supply pipes are planned with DN 63.

Therefore, according to the pressure loss table, this gives us an overall pressure loss for the feed and return pipes of around:

$$P_P = 4.3 \text{ mWs} / 100 \text{ m} \times 40 \text{ m} = 1.72 \text{ mWs} = 0.17 \text{ bar}$$

Each heat pump requires $2 \times 4 = 8$ moulded parts between the manifold to the heat pump. The pressure loss over the moulded parts is calculated as:

$$P_{MB} = 4.3 \text{ mWs} / 100 \text{ m} \times (8 \times 1.4 \text{ m}) = 0.48 \text{ mWs} = 0.05 \text{ bar}$$

$$P_{HPE} = \text{pressure loss over the heat pump evaporator is: } 0.27 \text{ bar}$$

For the cooling:

$$P_{HEC} = 20 \text{ kPa} = 0.2 \text{ bar}$$

Therefore, for each heat pump:

$$\Delta P = P_L + P_P + P_{MB} + P_{HPE} + P_{HEC}$$
$$= 0.46 \text{ bar} + 0.17 \text{ bar} + 0.05 \text{ bar} + 0.27 \text{ bar} + 0.2 \text{ bar}$$
$$= 1.15 \text{ bar}$$
$$= 11.5 \text{ mWs}$$

This gives us brine circulation pump SW 40.100 for the heat pumps:

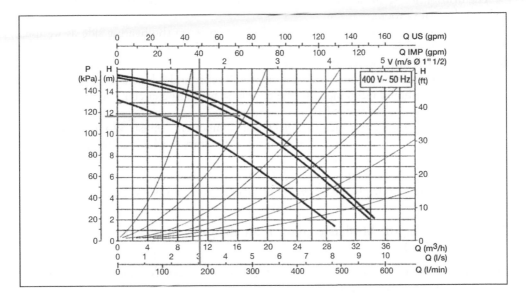

Figure 18.33 Selecting the brine circulation pump

This gives us the DAB brine circulation pump model BPH 150/250.50 T.
    This brine pump can be operated on level 2.
    The borehole heat exchangers could be drilled as follows:

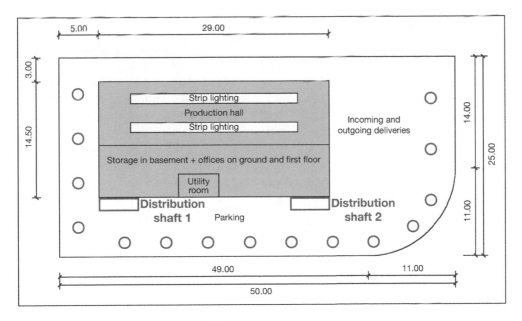

Figure 18.34 Positioning of the borehole heat exchangers

The borehole heat exchangers can be evenly distributed over the grounds.

The distribution shafts for the brine circuit manifolds can be positioned to the left and right of the utility room.

We use the same calculations for the charging pumps on the heating side as we used for the water-water heat pumps. However, we must remember to consider the larger nominal flow rates for the brine-water heat pumps.

As a water-water heat pump offers lower overall costs and a better level of efficiency, this is what you should recommend to the customer, where the circumstances permit and safe operation can be guaranteed.

Before completing planning, the electrical components must also be considered (i.e. cable cross-section, fuses, and the need for a soft starter, which is likely considering the large starting currents involved here).

This example was not simple, showing clearly that a heat pump system needs to be very carefully and thoroughly planned, because planning errors can be very expensive!

18.5 Calculating the bivalence point of an air-water heat pump (task 15.5)

First, you advise your customer, reminding him to also consider hot water generation (i.e. $P_{HW} = 4 \times 300$ W = 1.2 kW).

$$\rightarrow \quad P_{H'} = P_H + P_{HW} = 12 \text{ kW} + 1.2 \text{ kW} = 13.2 \text{ kW}$$

Table 18.9 Technical data for air-water heat pumps

Technical data – weight

Series Besst – Besst/r

Reference standard UNI-EN 14511:2004

Model		1/9	2/10	3/11	4/12	5/13	6/14	7/15	8/16
Heat output[1]	kW	6.8	8.3	11.0	15.0	19.9	22.2	28.0	37.2
Power output[1]	kW	1.74	2.11	2.81	3.61	4.28	4.83	6.48	8.44
Heat output[2]	kW	5.9	7.2	9.5	13.0	16.6	18.9	23.8	31.7
Power output[2]	kW	1.78	2.20	2.90	3.67	4.33	4.86	6.55	8.52
Heat output[3]	kW	5.0	6.2	8.0	10.8	13.5	15.4	19.2	25.8
Power output[3]	kW	1.79	2.26	2.97	3.73	4.36	4.88	6.66	8.63
Heat output[4]	kW	6.6	8.1	10.6	14.4	18.4	21.0	26.4	34.8
Power output[4]	kW	2.14	2.68	3.50	4.49	5.23	5.96	7.84	10.18
Heat output[5]	kW	5.6	7.1	9.3	12.6	15.9	18.1	22.6	30.0
Power output[5]	kW	2.16	2.77	3.60	4.60	5.26	5.99	8.00	10.33

1 External air temperature +7 °C Water 35 – 30 °C
2 External air temperature +0 °C Water 35 – 30 °C
3 External air temperature –7 °C Water 35 – 30 °C
4 External air temperature +7 °C Water 45 – 40 °C
5 External air temperature +0 °C Water 45 – 40 °C
6 External air temperature +30 °C Water 18 – 23 °C
7 External air temperature +35 °C Water 18 – 23 °C
8 External air temperature +30 °C Water 7 – 12 °C
9 External air temperature +35 °C Water 7 – 12 °C

You advise your customer and recommend a larger heat pump, at least model 4/12. (3) immediately shows us that auxiliary heating is required, even for this heat pump. Even the next largest heat pump, model 5/13, probably could not be operated in bivalent mode at −12 °C.

Now we need to determine the bivalence point of the two heat pump models 3/11 and 4/12.

First, for heat pump model 3/11:

$$P_{HP}(T) = a_0 + a_1 \times T$$

$$P_{HP}(T = 7\ °C) = 11.0\ kW = a_0 + a_1 \times 7\ °C \tag{1}$$

and

$$P_{HP}(T = -7\ °C) = 8.0\ kW = a_0 + a_1 \times (-7\ °C) \tag{2}$$

$$\rightarrow \quad a_0 = 8.0\ kW - a_1 \times (-7\ °C) \tag{3}$$

Equation (3) in (1):

$$P_{HP}(T = 7\ °C) = 11.0\ kW = a_0 + a_1 \times 7\ °C$$
$$= 8.0\ kW - a_1 \times (-7\ °C) + a_1 \times 7\ °C$$
$$= 8.0\ kW + a_1 \times 14\ °C$$

$$\rightarrow \quad 3.0\ kW = a_1 \times 14\ °C$$

$$\rightarrow \quad a_1 = 3.0\ kW\ /\ 14\ °C = 0.214\ kW/°C \tag{4}$$

Equation (4) in (3):

$$a_0 = 8.0\ kW - a_1 \times (-7\ °C)$$
$$= 8.0\ kW - 0.214\ kW/°C \times (-7\ °C)$$
$$= 8.0\ kW + 1.5\ kW$$
$$= 9.5\ kW$$

$$\rightarrow \quad P_{HP}(T) = 9.5\ kW + 0.214\ kW/°C \times T$$

Now we can calculate the output of the heat pump for the external reference temperature of −12 °C:

$$P_{HP}(T_N) = 9.5\ kW + 0.214\ kW/°C \times (-12\ °C)$$
$$= 9.5\ kW - 2.6\ kW$$
$$= 6.9\ kW$$

This gives us:

$$Q_{HP}\ /\ Q_N = 6.9\ kW\ /\ 13.2\ kW = 0.52$$

This means that under monoenergetic operation, the heat pump will cover up to 90 per cent of annual heat output. With a COP of around 3.2, power consumption increases to around:

90 per cent + 10 per cent × 3.2 = 122 per cent

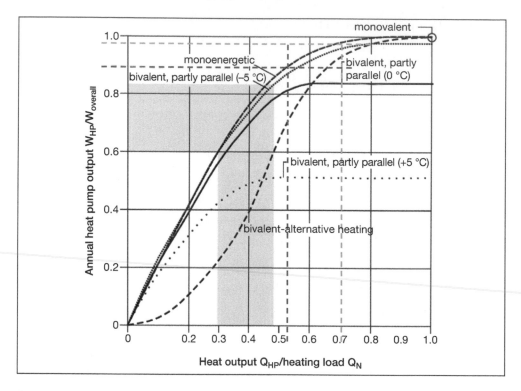

Figure 18.35 Determining the proportional heat output of air-water heat pumps

The next largest heat pump, model 4/12, gives the following equation:

$$P_{HP}(T) = 12.9 \text{ kW} + 0.3 \text{ kW/°C} \times T$$

At –12 °C, this gives a heat output of:

$$P_{HP}(T_N) = 12.9 \text{ kW} + 0.3 \text{ kW/°C} \times (-12 \text{ °C})$$

$$= 12.9 \text{ kW} - 3.6 \text{ kW}$$

$$= 9.3 \text{ kW}$$

For the heat pump:

$$Q_{HP} / Q_N = 9.3 \text{ kW} / 13.2 \text{ kW} = 0.7$$

Under monoenergetic operation, this heat pump will provide around 97 per cent of annual heat output and therefore operates in a significantly more economic and ecological manner.

18.6 Calculating various energy input factors (task 15.6)

The energy input factor is used to evaluate a building in ecological terms. It also reflects the economic efficiency of a heat pump system.

18.6.1 Calculating the energy input factor e_P for a detached house with a water-water heat pump for heating the building and generating domestic hot water

The following data are known for use in the calculation:

- Annual heat requirement: Q_H = 12,000 kWh/a
- Area to be heated: 230 m²
- Share of heat output for hot water generation: 20 per cent
- Underwater pump: 3.6 m³/h
- Groundwater level including drop: 9 m
- Charging pump output: 140 W

The energy input factor e_P is calculated as:

$$e_P = fp \: / \: \beta$$

$$fp = 2.6 - \text{for electrical heating} = \text{from the power grid}$$

The seasonal performance was calculated in Section 4.2.4, 'A comparison of the seasonal performance of the various heat pump systems':

$$\beta = 5.8$$

$$\rightarrow \quad e_P = 2.6 \: / \: 5.8 = 0.45$$

This calculation shows that a heat pump not only lowers energy costs, but also energy consumption!

18.6.2 Calculation of the energy input factor e_P of a detached house with a brine-water heat pump for heating the building and with electric hot water generation using an electrical heating element

The following data are known for use in the calculation:

- Annual heat requirement: Q_H = 12,000 kWh/a
- Area to be heated: 230 m²
- Share of heat output for hot water generation: 20 per cent
- Heating load of the building: 11.5 kW
- COP B0W35: 4.2

- COP B0W55: 2.5
- P_{HPH} = heat pump heat output for heating at B0W35: 12.5 kW
- P_{HPHW} = heat pump heat output for hot water generation at B0W55: 11.6 kW
- Brine circulation pump output: 160 W
- Charging pump output: 140 W

The energy input factor e_P is calculated as:

$$e_P = fp / \beta$$

fp = 2.6 – for electrical heating = from the power grid

Now we need to calculate the seasonal performance factor for this heat pump system. The seasonal performance factor is calculated as:

$$\beta = Q / E$$

where β = seasonal performance factor
Q = annual heat requirement for the building [kWh/a]
E = annual power consumption for all the heat pump system's electrical components [kWh/a]

The annual heat requirement is calculated as the sum of the annual heat output needed for heating and for generating hot water:

$$Q = Q_H + Q_{HW}$$

where Q_H = annual heat output for heating [kWh/a] = 12,000 kWh/a
Q_{HW} = annual heat output for hot water generation [kWh/a]
= 20 per cent of Q_H = 2,400 kWh/a

Annual power consumption for all electrical components in the heat pump system is calculated as:

$$E = E_{HP} + E_{PP}$$

where E_{HP} = annual energy consumption for the heat pump [kWh/a]
E_{PP} = annual energy consumption for the primary pump (brine pump) [kWh/a]

Annual power consumption for the heat pump is calculated as:

$$E_{HP} = Q_H / COP_H = Q_{HW} / COP_{HW}$$

where $COP_H = COP_{W10W30}$ for heating B0W30 = 1.23 × COP_{W10W35}
COP_{HW} = COP for hot water generation B0W55
→ E_{HP} = 12,000 kWh/a / (1.23 × 4.2) + 2,400 kWh/a / 2.5 = 3,282 kWh/a

The annual power consumption for the primary pump is calculated as:

$$E_{PP} = P_{PP} \times t_{PP}$$

where P_{PP} = electrical output of the primary pump (brine pump) [kW]
t_{PP} = annual operating time for the primary pump (brine pump) [h]

and

$$P_{PP} = P_L \times Q_{HP}$$

where P_L = pressure loss over the entire brine circuit [Pa] = [kg/ms²]
Q_{HP} = volume of brine flowing through the heat pump [m³/h]

In an initial approximation, we can assume:

$$t_{PP} = t$$

where t = system operating time [h]
$t_{PP} = t = Q_H / P_{HPH} + Q_H / P_{HPHW}$

where P_{HPH} = heat pump heat output for heating at BOW35 = 12.5 kw
P_{HPHW} = heat pump heat output for hot water generation at BOW55 = 11.6 kW

→ $t_{PP} = t$ = 12,000 kWh/a / 12.5 kw + 2,400 kWh/a / 11.6 kw

= 960 h/a + 207 h/a

= 1,167 h/a

The annual energy consumption for the brine pump is calculated as:

$$E_{PP} = P_{PP} \times t_{PP} = 160 \text{ W} \times 1,167 \text{ h/a} = 186,720 \text{ Wh/a} = 187 \text{ kWh/a}$$

The annual energy consumption for the secondary pump is calculated as:

→ E = E_{HP} + E_{PP} + E_{SP} = 3,282 kWh/a + 187 kWh/a = 3,470 kwh/a

→ β = Q / E = 14,400 kWh/a / 3,470 kWh/a = 4.15

→ e_p = fp / β = 2.6 / 4.15 = 0.63

19 Answers

Here we answer the questions listed in Chapter 16, 'Questions'.

1. **How does a heat pump work? (see Section 3.3, 'The technical refrigeration cycle and the function of a heat pump')**

In principle, heat pumps function according to the Carnot cycle. The gaseous refrigerant is compressed in the compressor. This strongly heats up the refrigerant. The refrigerant then passes into the condenser where it is converted into a liquid state. In changing state (from a gas into a liquid), the refrigerant gives off large quantities of heat. The heating water flowing through the condenser in the secondary circuit uses this released heat for heating and generating domestic hot water. The liquid refrigerant then flows through the injector, allowing it to expand and cool down significantly. Once in the evaporator, the refrigerant is heated and evaporates. In changing state (from a liquid into a gas), the refrigerant absorbs large quantities of heat. This heat is provided by the medium (water or brine) flowing through the evaporator in the primary circuit.

2. **Can a heat pump also be used to generate domestic hot water? (see Section 3.11, 'Boiler heat pumps', Section 3.12, 'Boiler heat pumps with heat recovery from the exhaust air', and Section 7.4.3, 'Hot water generation using the heat pump via a hot water tank')**

Of course, a heat pump can also be used to generate domestic hot water. Hot water should be generated using the heat pump because this is almost three times less expensive than heating water directly using electricity (e.g. with an electric heating element). It usually makes no sense to use an electric heating element to generate hot water directly if a heat pump is available.

3. **Is an additional form of heating required? Can a heat pump cover all the heating requirements? (see Chapter 5)**

When a heat pump system has been correctly planned and dimensioned, then no supplementary heating is required. An adequately sized heat pump provides enough heat for heating a building and for hot water generation. Using a supplementary form of heating, especially an electric heating element, has a significant, negative impact on the system's overall efficiency, and thus on the annual performance and energy input factors.

4. When is a heat pump worthwhile economically? Does it make economic sense for me?

Several factors determine if a heat pump makes economic sense:

1 It always makes economic sense to install a heat pump in a new building that has underfloor or wall surface heating (see Section 9.1, 'A simple water-water heat pump system'). That is why new builds should be planned to include surface heating systems.

2 Renovating an existing building and installing a heat pump can be particularly efficient because the energy savings will be greater in an existing building with little insulation than in a new build. However, the building does need underfloor heating, at least on the ground floor (heat rises).

3 At current energy prices, it is not economically worthwhile to install a heat pump in an existing building that needs to be heated with high flow temperatures (greater than 50 °C). If a heat pump is being installed, then the entire heating system will need to be overhauled, ideally with a water-water heat pump and low temperature radiators. Otherwise, it may make more sense to invest in energy-saving measures instead, including roof insulation and new windows, etc.

A heat pump system only makes economic sense if no direct electric heating is used.

5. Do you need an electric heating element with a heat pump?

Counter-question: do you need an electric heating element with oil-fired or gas heating systems? Then why would you need one with a heat pump? In principle, no! A heat pump system should be designed to operate in monovalent (not just monoenergetic!) mode. Then no additional heat source will be necessary. If a customer wishes to have an electric heating element, then it should only be used as a fallback in the case of emergency.

6. What are the alternatives to heat pumps?

There are plenty of alternatives – but are they real alternatives? As already explained, in principle, all new builds should be equipped with a heat pump. For ecological reasons, wood or pellet boilers are often recommended, because their closed CO_2 cycles place no additional burden on the environment. This can also be achieved with a heat pump if the operator uses the green electricity offered by their local utility. (Utilities in Germany are only permitted to sell as much green electricity as they actually produce.)

Where there are high flow temperatures, a heat pump makes no sense, and here the search for alternatives such as wood or pellet boilers is recommended. However, increasing discussion about carcinogenic substances in pellet dust and dust pollution is making these alternatives less attractive.

When power demand is high (e.g. for small businesses), then a combined heat and power unit (CHP) may be a better alternative.

7. What is a bivalence point?

Heat pumps operated in bivalent mode cover heating demand up to the so-called bivalence point. Thus the bivalence point is effectively the external temperature at which a second heating system kicks in.

It is worth pointing out here that, in principle, it is better to operate a heat pump in monovalent mode. The question of whether bivalent mode is possible usually arises when renovating a building with an existing heating system. It is certainly technically possible to operate a heat pump in bivalent mode, but it makes little economic sense: bivalent mode means two heating systems – the existing one and the new heat pump. Both need to be serviced and maintained, and this increases costs. And what happens when the 'old' system dies? It will have to be repaired or replaced. Therefore, it is clear that this arrangement makes little economic sense.

8. What are the key factors that need to be taken into account when considering existing buildings?

The following must be taken into account when renovating existing building stock:

1   First, the flow temperatures needed for heating must be determined. If these temperatures are too high, then a heat pump will not be suitable and alternatives must be considered (see Section 4.3, 'Seasonal performance factor').
2   Where the building has underfloor heating and radiators, then the maximum required flow temperatures must be determined. If these are more than 5 °C apart (maximum 10 °C), then two separate buffer tanks are recommended so that the underfloor heating can be operated using the lowest possible flow temperatures and with the best possible COP.
3   The heat output of the existing boiler should not be used as the reference value when determining required heat output because the existing boiler will often be larger than necessary. It is a good idea to ask a building energy consultant or planning office to determine the necessary heat output.

9. How long will a heat pump last?

How old is your fridge, or the one your parents or grandparents have? A heat pump is a refrigerator. It works like a fridge and can last just as long. What is important is that it is correctly dimensioned, and is operated using a buffer tank.

10. How high are the life cycle costs?

No oil or gas costs, no maintenance costs, no chimney sweep, only the cost of electricity. However, these electricity costs are limited because electricity is primarily used to run the compressor and the pumps. Electricity is only an auxiliary form of power, enabling the free geothermal energy to be used.

11. How often must a heat pump be serviced?

I recommend dusting the heat pump once a year. The heat pump requires no other regular service. To ensure that the heating system requires no maintenance, the hot water tank should be fitted with a sacrificial anode. Because the system needs no regular service, customers often forget that the magnesium anode needs servicing, and this can lead to corrosion in the hot water tank.

It is important to ensure that underfloor heating is absolutely oxygen-tight to exclude the possibility of corrosion in the buffer tank (steel).

12. Why have a buffer tank? (see Section 7.3, 'The buffer tank')

A buffer tank has several functions:

1 As a hydraulic separator, so that the buffer tank is charged with enough water to ensure the optimal flow rate needed by the heat pump, and that the heating system is fed with the ideal quantity of heat (i.e. flow rate) subject to changes in the weather.
2 So that the heat pump is turned on and off as little as possible, therefore prolonging its life.
3 As a central heat storage unit (e.g. when other energy generators such as solar systems and wood stoves are also connected to the heating system).

13. What faults can be expected? (see Section 13, 'Common heat pump errors')

If properly planned, installed and operated, a heat pump, like all refrigeration units, is not susceptible to faults.

The usual faults are described in Chapter 13.

14. What is a low pressure fault? What causes them? (see Section 13.5, 'Error messages and their possible causes')

A low pressure fault occurs when the refrigerant on the primary side (water or brine) is overcooled. This is caused by too little heat being transferred from the heat source, either because there is too little of the transfer medium (water) or the brine is too cool (borehole or ground heat exchangers are too small).

Warning: low pressure faults can lead to water-water heat pumps freezing!

15. What is a high pressure fault? What causes them? (see Section 13.5, 'Error messages and their possible causes')

A high pressure fault occurs when there is too little condensation in the condenser. This is caused by too little heat being transferred to the secondary heating circuit, and is usually because the charging pump is too small.

16. What is an evaporator? (see Section 3.2, 'Structure of a heat pump and its components', and Section 3.3, 'The technical refrigeration cycle and the function of a heat pump')

The evaporator is the heat exchanger on the heat source side (water or brine). Here, the refrigerant evaporates and becomes a gas, absorbing large quantities of energy.

17. What is a condenser? (see Section 3.2, 'Structure of a heat pump and its components', and Section 3.3, 'The technical refrigeration cycle and the function of a heat pump')

The condenser is the heat exchanger on the heating side. Here, the refrigerant condenses and becomes liquid, releasing large quantities of energy that is used for heating.

18. **What is a compressor? (see Section 3.2, 'Structure of a heat pump and its components', and Section 3.3, 'The technical refrigeration cycle and the function of a heat pump')**

The compressor compresses the refrigerant, causing it to heat up strongly.

19. **What is a dryer? What is it used for? (see Section 3.3, 'The technical refrigeration cycle and the function of a heat pump')**

Despite extracting air from the refrigerant cycle when installing a heat pump, residual damp always remains. This is removed by the dryer.

20. **What is the observation glass for? (see Section 3.3, 'The technical refrigeration cycle and the function of a heat pump')**

By looking through the observation glass, the service technician can see the operating conditions inside a heat pump.

21. **What is a refrigerant collector? What is its role? (see Section 3.3, 'The technical refrigeration cycle and the function of a heat pump')**

The collector is installed behind the condenser. It collects the refrigerant leaving the condenser and feeds it into the injector in a liquid state.

22. **What is an injector? (see Section 3.3, 'The technical refrigeration cycle and the function of a heat pump')**

Once passed through the injector, the refrigerant expands and cools down strongly. The cooled refrigerant is now ready to absorb energy in the evaporator.

23. **What is bivalent heat pump operation?**

The term comes from the Latin 'bi', meaning two, indicating that a bivalent heat pump has an extra means of heating. This could be an additional gas or oil-fired boiler, or an electric heating element. As we have already established, a heat pump should always be operated in monovalent – not bivalent – mode.

24. **What is monovalent heat pump operation?**

This is also derived from the Latin: 'mono' means one or alone. A heat pump operating in monovalent mode is the opposite to one operating in bivalent mode. Monovalent mode means the heat pump uses geothermal energy from the ground or the air, and auxiliary sources of energy (usually electricity) are only used to operate the compressor, the pumps and controllers. Electricity is certainly not used for an electric heating element, either to provide additional heating or to generate hot water.

25. **What is monoenergetic heat pump operation?**

Monoenergetic operation for a heat pump system means that the system is driven only by a single source of energy, usually electricity. But be careful: here, the electricity is also used directly for heating, via an electric heating element.

26. Why should a heat pump be operated in monovalent mode?

For the reasons given above, in principle a heat pump should be operated in monovalent mode. Extra boilers involve greater servicing and maintenance costs. Monovalent operation offers the most efficient method of operation and the shortest amortisation times.

Index

absorbers 134–7
absorption heat pump 44–51
active cooling 41–2
adsorption 48
aerothermal energy 35
air-air pump 51–2
air conditioning 51–3, 100
air-water pump 35–8, 105, 137–49, 255; COP 60–1; modes 149–52; seasonal performance 66–9, 74; starting up 224
ammonia 46
amortisation 212–15; see also economics
answers 318–23
applications 52–4
architect 79–82
Austria 5
authorisation 154–6

bivalence 79, 255, 312–15; air-water 139–52
blocking periods 84–7
boiler 38–9, 78
borehole heat exchanger 31–2, 34; errors 227; planning 102–5, 115–20, 122–6
brazed heat exchanger 110
brine circulation 177–9
brine-water pump 30–3, 102–5; COP 59–60; exercises 252–3, 276–94; hazards 160–1; planning 114–15, 122–6, 129–32, 152; seasonal performance 64–6, 73–4; system 204–5, 224
buffer tank 187–90

camping 52
charging pump 179–80, 182–7
chemicals 26–7, 46
climate change 2, 5, 26–7, 220
CO_2 borehole heat exchanger 164–5
coefficient of performance (COP) 55–65, 67, 71–4

commercial use 99–100
compact unit 37–8, 79–80, 206–8
components 12–7, 206
compressor 13, 197
condenser 14
construction 81–2
consultant 79–82, 210
consumption see power consumption
controller 17, 20–1, 42–3, 198, 230
cooling 28, 39–43, 126, 204–5; capacity 114–15
COP see coefficient of performance
cut-out switch 16–17

data sheets 233–48
desorption 49
dimensioning 79–80, 138, 174; charging pump 182–7; compact unit 206–8; piping 197
DIN 8901 159, 219–20;
DIN EN 12828 163
DIN EN 12831 87, 96, 106, 137
DIN EN 14511 55, 59, 61, 65, 220
direct cooling 41–2
direct evaporator 34–5
discharge wells 111–12
ditch heat exchangers 134
Dorsten-Wulfen 232, 261–3
dryer 16
drying 224–5
dry running 173, 229

ecology see environment
economics 6–9, 92, 100, 162–3, 202, 209–16
efficiency 80, 212; COP 55–65, 67, 71–4; see also seasonal performance
EHPA see European heat pump association
electronics 197–8; electrician 82; heating element 192; installation 154

energy: aerothermal 35; consumption 66, 69, 71–3; costs 2, 6–9, 88, 152, 209–15; efficiency 187; fences 134–5; input factor 76–7, 256, 315–17; mats 132–3; regulations 217–20; renewable 5, 77, 218; *see also* output
engineer 78, 226
environment 2–6, 54, 77–8, 231–2; Geo-Protector 30, 162; legal regulations 217–21
errors 226–30
European heat pump association (EHPA) 70
European standards 219–20; *see also* DIN; DIN EN
evaporator 15, 17, 34–5, 167
examples 259–63
exercises: bivalence 255; brine-water pump 252–3; data sheets 233–48; energy input 256; large pump 254–5; water-water pump 249–52; *see also* solutions
existing buildings 87–93
expansion valve 14–15

flow: monitor 173, 224, 229; rate 167–71, 180, 226; temperature 230; *see also* COP
free cooling 40
funding 156–7
fuses 197

garage 203
gas-driven heat pump 43–51
Geo-Protector 159–64
geology 108–9, 124–5
geothermal: baskets 133; energy 5, 9, 105, 231–2
Germany 5, 26–7, 154–6, 158, 209; Dorsten-Wulfen 232, 261–3; legal regulations 217–21
global warming *see* climate change
ground heat exchanger 32–4, 127–32, 227
groundwater 28, 109–12, 158–65, 220–1
guidelines 199–201

hazards 160–1
heat exchangers 31–4, 81, 102–3, 105; brazed 110; brine circulation 177–9; CO_2 164–5; ditch 134; errors 227; hot water 193–4; *see also* borehole heat exchanger; ground heat exchanger
heating systems 157
heat output *see* output

heat pump 10–12, 52–4, 91–3, 231–2; air-air 51–2; boiler 38–9; controller 20–1; cooling 39–43; direct evaporator 34–5; errors 226–30; examples 259–63; existing buildings 88–91; gas-driven 43–51; refrigeration 13–9, 22–7; structure 12, 19–20; system 199–208, 222–5; *see also* air-water pump; brine-water pump; hydraulics; output; planning; water-water pump
heat source 78, 82
high pressure 229
hot gas 18–19
hot water 191–7, 228
housing 202–3, 232; existing 87–93; new build 82–7; output 96–8, 101–6
hydraulics 166–7; buffer tank 187–90; charging pump 179–80, 182–7; electronics 197–8; errors 227; hot water 191–7; primary pump 167, 170–9
hydrology 108–9
hydrothermal energy 27

industrial halls 93–6
installation 81–2, 228; electrical 154; heat exchanger 122; *see also* dimensioning
instantaneous water heater 191
insulation 106
intake wells 111–12
iron 28, 60, 91, 107–10, 112, 232

Kreditanstalt für Wiederaufbau (KfW) (Germany's Reconstruction and Loan Corporation) 106

lake absorber 135–7
large pumps 101–5, 254–5, 295–312
legal regulations 158–9, 217–21; *see also* DIN; DIN EN; VDI
low pressure 229

maintenance 207
manganese 28, 60, 91, 107–10, 112, 232
monitors 173, 176, 198
monoenergetic 150–1
monovalence 149
motor protection switch 172
municipalities 213

new build 82–7
night mode 230
noise 105

observation glass 16
operator faults 230
outbuildings 81–2
output 78–9; commercial use 99–100; COP
    55–61; existing buildings 87–93; housing
    96–8, 101–6; industrial halls 93–6; new
    build 82–7; seasonal performance 62–70;
    SPF 70–5; swimming pool 98–9

passive houses 106
pellet boilers 213–15
performance: seasonal 62–70; seasonal
    performance factor (SPF) 70–5; *see also*
    COP
piping 170–6, 178–9, 197
planning 78–9, 199–201; absorbers 134–7; air-
    water 137–52; applications 154–6;
    architect 79–82; brine-water 114–20,
    122–6, 129–32, 252–3; energy mats 132–3;
    funding 156–7; heat exchanger 115–20,
    122–9, 134; quality 152, 154; water-water
    106–13, 249–52; *see also* hydraulics; output
power: consumption 66, 69, 71–3; supply
    197–8; *see also* electronics; energy
pressure 229; fault 29; loss 167–71, 175,
    177–80, 182–4
primary energy factor 76
primary pump 167, 170–9, 223
probe 17
pump *see* heat pump; hydraulics

quality 152, 154
questions 257–8; answers to 318–23

radiators 88–91
refrigerants 16, 23–7, 46,
refrigeration cycle 13–19, 22
regions 213
renewable energy 5, 77, 218
repairs *see* maintenance
residential areas *see* housing
resources 6–7
reversible heat pump 41–2
river absorber 135–7
roof absorber 134
Ruskin, J. 216

safety chain 229
seasonal performance 62–70; factor 70–5, 90

secondary pump *see* charging pump
sensors 198, 228
soft starter 197
solar power 5, 48–51, 104–5, 195–6, 203, 205
solid absorbers 134
solutions: bivalence 312–15; brine
    water/boreholes 276–84; brine
    water/ground 284–94; energy input
    315–17; large pump 295–312;
    water-water pump 264–75
space *see* dimensioning
split unit 38
stand-alone units 206
starting up 222–5
statutory provisions 158–9; *see also* DIN; DIN
    EN; VDI
stove 188, 202
structure 12, 19–20
surface heating 157
Sweden 5
swimming pools 80–1, 98–9, 203
Switzerland 5
system 202–8, 222–5; guidelines 199–201;
    separation 29–30

temperature glide 24
thermal compressor 44–5
thermal expansion valve 14–15
Tichelmann, A. 127

underfloor heating 88–90, 96, 187–8
underwater pump 167, 170–6, 223–4

Vaillant 47, 51
VDI 4640 119, 128, 159, 162, 218–9
VDI 4650 61, 74–5, 117, 219
von Linde, C. 10

warming 224–5
water: hot 191–7; quality 109–10; supply
    172–6; *see also* groundwater
water-water pump 27–30, 40, 101–2, 106–13;
    COP 58–60; planning 152, 249–52,
    264–75; seasonal performance 62–4, 71–3;
    system 203–4, 223–4
wells 108–9, 111–12, 226
Wohnlich, S. 158

zeolite 47–51